Mountains in the Greenhouse

Donald McKenzie

Mountains in the Greenhouse

Climate Change and the Mountains
of the Western U.S.A.

 Springer

Donald McKenzie
School of Environmental and Forest Sciences
University of Washington
Seattle, WA, USA

ISBN 978-3-030-42431-2   ISBN 978-3-030-42432-9   (eBook)
https://doi.org/10.1007/978-3-030-42432-9

This Springer imprint is published by the registered company Springer Nature Switzerland AG
The registered company address is: Gewerbestrasse 11, 6330 Cham, Switzerland

*To **Doug McKenzie**: mountaineer extraordinaire, astute and thoughtful student of climate-change science, and a best friend for more than 50 years.*

Preface

The idea for this book began with the lead scientists in a research consortium called the Western Mountain Initiative. The WMI was supported by federal agencies and academic sources and addressed the topic of the book, climate change, and the mountain ecosystems of the American West, with field studies and computer models. Some of us were based in National Parks where we had worked for years, predating the WMI. These include Sequoia National Park in the Sierra Nevada, Glacier National Park in the Northern Rocky Mountains, North Cascades National Park in the Cascade Range, Rocky Mountain National Park in the Central Rocky Mountains, and Bandelier National Monument in the Southern Rocky Mountains. We synthesized what we learned from the field studies in the Parks into West-wide models, and we wrote hundreds of scientific papers and agency reports. Given that the American taxpayers financed much of this work, we wanted to translate and distribute it back to them, as widely as possible, in a way that was comprehensible but not diluted.

A book that distills the work of many can be written by many, or a few, or a single author. In the single-author case, the reader may see the process of construction more clearly than if it were the product of many minds. This is important because I hope that you can follow the story line here in a particular way. My point is not to convince anyone that global warming and its effects are real. That would require a different educational strategy that I leave to others who excel at it. If you can understand the processes by which we infer what has happened already as mountain ecosystems respond to a warming climate, and by which we estimate what will happen in the future, then I have succeeded.

Much as in a theatrical production, you will see the stage (the mountains), meet the players, and observe how they interact. Unlike the theater, however, we have no poetic license; we are in the realm of science, in which (to paraphrase biologist Thomas Huxley) "a beautiful theory [or a script, in this case] may be slain by an ugly fact". The plot of our story is constrained by the research of the WMI and many others in the West and around the world.

Readers who took advanced math courses surely encountered textbooks with titles that began "A First course in..." or "Introduction to...". The author then explains how the book is entirely self-contained, so you do not need anything else to supplement it. In these books, more often than not I would be lost by page 5. With that experience in mind, I am going to tell you now that this book is self-contained. Planned redundancies provide different filters and angles on key concepts and subjects. Different ways to read it take advantage of the redundancy, and these ways can be mixed. One way is to read in order through the text, saving the notes for later. The text is an unbroken story. Chapter 1 is introductory, and its last section on "concepts that you need" should be kept handy. These are the longest definitions in the book, and you will need them as you go along but should avoid bogging down in them to start, when you will have less context for them. There is also a glossary with 2–3 sentence definitions. Words that are italicized in the text but not defined immediately will be in the glossary or the terms in Chapter 1. Chapter 2 gives the setting, with lots of geography but also the key ecological aspects of different mountain ranges. Chapters 3–6 are sequential, with each building on the previous, so they are best read in that order even if you skip around elsewhere. Chapters 7–9 can be read in any order.

Another way to read is to look at some or all of the notes as you go. This will be easier for those with the ebook version than the print version. I parsed the narrative into two levels so that you can hold the larger view (the proverbial "forest") and not be distracted by its "trees", but you can drop into their midst at any time. Some notes give technical details; others are anecdotal. If the text has paraphrased any published work, that work will be in the associated note(s). Italicized words in the text will be defined in the associated note(s) or parenthetically in the text.

A third way to read is the "Congressional staffer" or "twitter" method, for which I provide a means. At the end of most chapters, there are sections that summarize the expectations about that chapter's topics for each of the mountain ranges we cover. As a quick dip, or a review, these can be understood without in-depth reading of the chapters.

Chapter 8 will be the most difficult for most readers. If you are confused, you are in good company. Cutting-edge researchers (and your author) are still trying to "tame" some of the wildly varying observations that we are making. Because of the topics themselves, I have written this chapter similarly to how I have written technical papers on these same subjects. No equations here, but lots of concepts, and more of my personal views than in most other chapters.

Chapter 9 also mixes scientific consensus with the personal. I end by returning to the concepts of "wildness" and "wilderness" that drew me to our mountains in the first place. I invite you to go there too, as much as you are willing.

Seattle, WA, USA   Donald McKenzie

Acknowledgments

Many have helped me to complete this book, in large and small ways, directly and indirectly. Dave Peterson was a co-lead in the initial stages, which covered years. The overall structure and the title were as much his ideas as mine. Patti Loesche, my brilliant editor from concepts to commas, and Zen Master of writing, wore an array of hats and found ways to improve this work under each one. Rob Norheim crafted the maps in Chapters 2 and 6, and for years has been my go-to source for when and how to use cartography. My other colleagues (besides Dave P) in the Western Mountain Initiative added momentum early on and feedback throughout: Naomi Tague, Nate Stephenson, Dennis Ojima, Jeff Hicke, Andrew Fountain, Dan Fagre, Jill Baron, Craig Allen. My graduate students and postdocs gave me inspiration, and hope for the future, and taught me far more than I ever taught them: Alina Cansler, Lara-Karena Kellogg, Maureen Kennedy, Karen Kopper, Jill Nakawatase, Erica Newman, Crystal Raymond, Tessia Robbins, Natasha Stavros. Nineteen family members, friends, and colleagues contributed Western Mountain photos; their names appear with their artwork in the text. More photos are courtesy of Conservation Northwest, People and Carnivores, the Southern Utah Wilderness Alliance, the National Park Service, the US EPA, and the USGS. Over the years, many friends and colleagues have kept me honest in science, and writing, and made me better at everything I do: Uma Shankar, Isabel Ramírez, Roger Ottmar, Carol Miller, Connie Millar, Theresa May, Jeremy Littell, Bob Keane, Amy Hessl, Charlie Halpern, Nancy French, Don Falk, David Wolpert, Monica Turner, Tom Swetnam, Chris Stalling, Amy Snover, Erica Smithwick, Susan O'Neill, Ed Miles, Sim Larkin, Emily Heyerdahl, Ze'ev Gedalof, Janet Franklin, Sam Cushman. Hundreds of public servants—lower in the hierarchy than political appointees—have helped the Western Mountains to keep more of their character than they might have. These folks work for the federal government, Indian tribes, states, counties, and cities. I thank my two agents at Springer for engaging me to write the book I always wanted to write and for making the whole process remarkably easy: Sher Saini and Margaret Deignan. Last and farthest from least, I thank Susan, Kira, Kyle, and Murphy for being a family who are way beyond amazing.

Contents

Chapter 1
Introduction: What Persists, What Changes

Sky, rock, sunlight, snow
The eternal wind rising
Follow the faint trail

Over a human lifetime, mountains themselves are essentially permanent. With the obvious exception of volcanoes, significant changes in landforms themselves happen on *geological time scales*: millions of years. On the land surface, however, water flows, plants grow, animals live and die, and a host of other processes occur over periods from seconds to years. We can measure the time scales on which climate change occurs; right now, faster than ever,[1] but still over decades and centuries. Its effects on those processes on the land surface can happen more quickly, however. Those effects, and the time scales involved, are the subject of this book.

My first experience of mountains as a child was their verticality. Placed on a pair of boards that slid on the snow, I soon learned how it felt to flow, but also to tumble. Gravity is the prime mover on mountain landscapes. Its most persistent and visible function is to move water from mountains to everywhere else, sometimes flowing, sometimes tumbling as water, snow, or ice. This verticality also produces a unique ecological feature: everything changes with elevation. Higher up, it's colder, often wetter, often windier, and at a certain point, there is too much of one or more of these to sustain living things, whether plants or animals.[2] Mountains have been called "water towers of the world", because gravity draws from them more than half of all the water used by humans.[3] These human populations rely on an immense complex infrastructure of water management, which itself is dependent on the

---

[1] As far as we know, that is, over the period of Earth's history for which we have figured it out. See Chapter 3 for details.

[2] Yes, there are other living things besides plants and animals, some of which can survive harsher conditions. My biological focus will be plants and animals.

[3] A good synopsis of this function is at http://www.fao.org/fileadmin/templates/mountain_partnership/doc/POLICY_BRIEFS/SDGs_and_mountains_water_EN.pdf.

© Springer Nature Switzerland AG 2020
D. McKenzie, *Mountains in the Greenhouse*,
https://doi.org/10.1007/978-3-030-42432-9_1

amount and timing of the delivery of water. Upstream of most humans, a different complex structure of snow, ice, and waterways nourishes non-human populations.

Mountains are climate- and weather-makers; both climate and weather in the mountains are different and much more variable than on the flatlands. This comes from *topography*, the shapes and sizes of *landforms*, the physical features of a landscape. Topography channels the wind, blocks the movement of moisture carried by air currents, and interacts with the atmosphere to change the air temperature, sometimes in unexpected ways.

Why These Mountains?

Our setting for this book is the mountains of the American West, which I will dub the "Western Mountains" hereafter. The map shows our full "domain", from the Pacific Ocean to the Great Plains and from the Canadian to the Mexico border (Fig. 1.1). The western and eastern boundaries are obvious. They are where the mountains end, at an ocean and a prairie. The northern and southern boundaries are only political. The only things that change abruptly at these borders are the jurisdictions. If we drew lines by *ecosystems*[4] instead of countries, our borders would look very different and there would be no straight lines. These "cutoffs" are ecologically arbitrary but serve to limit our scope without invalidating the story told by the mountains within them. Some of that story is already familiar to U.S. readers, who may live in the mountains, visited national parks, wilderness areas, or other less protected landscapes, or even have just driven or ridden across the West. I shall ask for much thoughtful attention in the pages ahead, and that may be easier for those who are somewhat to very familiar with the Western Mountains. More practically, these are the mountains that I know, having stayed (for better or worse) within these four borders for most of my own research and recreation.

We have already learned a lot about the effects of climate change by studying the Western Mountains. Some of us[5] see them as a grand "natural experiment". An experiment needs a "lab" and *controls* (ways of minimizing the number of things going on at once). Our lab is one of the most diverse on Earth, in that from west to east and north to south across the Western Mountains, there is more ecological variation than almost anywhere else.[6] As for control, there is just enough. We have enough access to the Western Mountains to study them, but not enough to confound

---

[4] I define ecosystem here as some collection of biological and physical elements that interact in ways that we can observe. For most of this book, these elements will be associated with tangible places, such as forests, meadows, or rivers. "Mountain ecosystems" will comprise all of them.

[5] For example, my colleagues and I in the Western Mountain Initiative, which I mentioned in the preface. But we are not the only ones.

[6] I don't mean "biodiversity" *per se*, for which the Amazon and other rainforests would be the winners. Rather it is all the players in our story: climate and weather, topography, rivers, forests, wildlife, fish. Read on.

Fig. 1.1 The Western Mountains, in global context, with the names of ranges that are discussed in this book. Map by Robert Norheim

the signals of climate change by human influences. This is not to say that there aren't any of those. The West overall has huge cities, dams and mines, and other "contaminants" to a natural experiment, but the climate-change signal is already evident, and will become more so.

Interesting provocative experiments are often lamented for their brevity. For example, we will know much more about how well a new cancer drug works,[7] or how a forest will recover after a wildfire, if we can study the process for 20 years instead of 3–5 years. These *designed* (as opposed to "natural") experiments are limited by practical concerns, such as how long research grants last, or how long it takes a graduate student to complete a degree. In contrast, our natural experiment, observing the effects of warming climate on ecosystems of the Western Mountains,

---

[7] This often means comparing how many people who took the drug either were free from cancer or just still alive at the end of the experiment vs. those same numbers for those who thought they were taking the drug but were given a fake ("placebo").

is limited by the pace of change. At some point will things be so different that we just can't predict anything meaningful about them? Yes, but no one knows exactly when that will happen, and as we shall see, it could be at different times in different regions. To give this context, however, I shall predict that it will happen on some Western-Mountain landscapes[8] within the lifetimes of some readers of this book. So you are not reading about a distant future here, but about the present and the near future.

Where Are We Going?

The setting for this natural experiment that we are observing has two elements, one of which persists throughout,[9] the mountains themselves, and one that changes constantly, the climate. Chapter 2 introduces the *persisting* mountains,[10] verbally and visually, with their similarities and contrasts. How large, how steep, how high, how continuous, how wet, how hot. What vegetation, what human history, what current human use. Chapter 3 introduces the *changing* climate. How much, how fast, where, how do we know, and what do we know better or worse.[11]

The players on this stage are many, and I have chosen not to cover them exhaustively, for practical reasons but also from expertise and interest. The subjects I have chosen for our natural experiment are water, vegetation, *disturbance*[12] (anything that brings an abrupt change to an ecosystem), and animals. The first three appear "on stage" in succession, reflecting not some natural hierarchy of physical and ecological processes but the order in which we overlay them as we build up knowledge of mountain ecosystems. Chapter 4 discusses how water and its frozen equivalents will change in a warming climate. How much more or less, where more or less, and especially when. Chapter 5 highlights trees and forests. Will they persist or disappear, where, and when. It also brings in the rest of the vegetation as carbon, the currency of many studies of climate change and policies to offset it. Chapter 6 is mostly about what *disturbances* do to vegetation, focusing on wildfire and insect outbreaks. Remove it, exclude it, change it. Chapter 7 switches kingdoms, from plant to animal, and asks what will happen to organisms that move in a warming climate. Will they move with the climate, do they need to, and what could stop them.

---

[8] The most vulnerable ones. By the end of your read, you should have a good idea which ones these are.

[9] With the exception of the aforementioned (but rare) volcanoes.

[10] Many of you may need even less introduction to the Western Mountains than I thought I might need when I started writing. This chapter focuses mostly, but not entirely, on their features of most relevance for our story. I hope that you may learn something new, even about your favorite range.

[11] But not "why you should believe it". As I said in the Preface, those seekers will have to look elsewhere.

[12] See Chapter 6 for a fuller definition.

Mountain ecology is complex, even without climate change, and few of the interesting questions that we ask are easily answered. I leave the (even) harder questions for Chapter 8. How do all these players interact, how incomplete is our knowledge, how far and how fast might things go off the rails, and finally which mountain ecosystems are vulnerable, and how vulnerable are they. When you get to Chapter 8, take a pause to let the previous pages settle in, and skip over things at first read if they seem dense.

In Chapter 9, I write explicitly about how humans fit in to the story. What do we need from the Western Mountains, what do we do about climate change, and what awaits those of us for whom the Western Mountains are important. How much will persist, and how much will change. On that last question, there are some broader elements that persist globally, and as we move ahead, their persistence will inform the question more specifically when applied to our setting, our players, and their interactions.

What Doesn't Change in What Changes

Mountain ecosystems change constantly, even in a stable climate. Sometimes the relationship between cause "A" and effect "B" will be different from one that was observed and verified earlier, in a different season, a different decade, or in a similar experiment or field study in a different place. For example, a tree species at the dry or warm edge of its range may thrive best in the wettest part of its local environment, whereas at the cold or wet edge it will thrive in the driest part. In a rapidly changing climate, such relationships are variable in both space and time, and in a relatively short time[13] may seem ephemeral, or just wrong.[14] The topics in this book will be rooted in a few "stabilizing" concepts: (1) ice versus water, (2) evolution, (3) movement, (4) interaction. #1 will be constant unless the laws of physics change. #3 and #4 will continue as long as there are biological organisms. #2 can be considered constant for the lifetime of relevance of this book,[15] or longer.

---

[13] For example, the time between when a book is written and when it is published.

[14] Of course, they may also be wrong because the science has improved in the interim, or because the research was conducted poorly. I will not address these issues here.

[15] Roughly, to the 2060s–2080s, beyond which my personal view is that at the rate of increase in global temperatures (fast) and of the response of societies (slow), the predictability of (Western USA) mountain ecology is effectively zero. The biologically informed reader could argue that some (micro)organisms evolve almost as fast as we breathe. Indeed, but this is not in response to warming climate, so I am ignoring it in this book.

Ice or Water, Snow or Rain

We can count on the stability of the freezing (melting) point of water (ice). Although it varies when moving versus still, and depending on dissolved materials, it does not vary with climate change. So as the climate warms, there is less ice, and more water; less snow, and more rain.[16] This dynamic can be counted on, underlying other complex phenomena that may be observed.[17]

Evolution

In the scientific literature you will see the expressions "geological time" (or the "geological time scale"), "evolutionary time", and "ecological time". These refer to the average or characteristic lengths of dominant processes, whether physical or biological. For example, in the geologic time frame, the Mesozoic Era spans the period between the greatest prehistoric extinction (the end of the Permian Period ~ 220 million years ago) and the extinction that killed the dinosaurs (end of the Cretaceous Period ~ 66 million years ago). In the context of the multi-million-year periods, evolution proceeds in almost a blur, but in the context (ours) of decades, evolutionary change is slow. For our purposes, evolution is a *constraint* (something that "holds back" a process) on biological responses to a rapidly warming climate, particularly for large organisms.[18] Basically, our climate is changing faster than (most) plant and animal species can evolve.

Movement

Organisms *disperse*, or *migrate*, or both, in response to factors more immediate and local than global temperature change. In turn, these factors respond to climate. Dispersal usually refers to permanent movement away from a starting point, for example, a parent tree for a seed, or the natal pack for a wolf. Migration is usual a

---

[16] A colleague of mine once reminded visiting agency officials and journalists that "Glaciers don't vote Democrat or Republican; they're just ice", regarding the shrinking and disappearance of glaciers across the West.

[17] Just as no matter what else is going on, what you observe won't violate the law of universal gravitation or the Second Law of Thermodynamics. If it does, check your algebra.

[18] Yes, bacteria, viruses, and cancer clones evolve much more rapidly than larger organisms, to the detriment of our health and our medical treatments and the health and survival of other large organisms. I will treat the rapid changes in micro-organisms as a constant factor here, while conceding that there are many subtleties in evolutionary change in response to changing climate.

semi-annual event (there and back), moving to the same location each year.[19] The constraints on dispersal and migration will have a big role in our exposition of responses to climate change, and I will consider them to be stable. For example, roads are a barrier to movement of large animals; that will remain true even if the local climate near a highway is changing quickly.

Interactions

None of the processes I will discuss happens in isolation. Interactions will persist in a warming climate, while their nature and strength change. Interactions are either *synergistic* (the outcome is "more" [+] than if they happened separately) or *antagonistic* (the outcome is "less" [−] than if they happened separately). In most cases of interest, but not all, interactions will keep the same "sign". Either way, they are a *sine qua non* of ecological science, whether the climate is changing or not.

Concepts and Terms You Should Know

Variable

A variable is something we care about that can change, such as wind speed, air temperature, or a tree's diameter or its annual growth rate. Often variables interact and affect other variables of interest, as air temperature (averaged over a year, or a summer) could affect a tree's growth rate. Models of climate change have to predict the values of up to thousands of variables. In the formula for the area of a circle, $A = \pi r^2$, A and r are variables, and π is a constant, whose value we know and doesn't change.

Parameter

A parameter is a number that "should" have some value, but we don't necessarily know what that value is. This is different from a variable, whose value can change, or a constant. Mathematical models and computer simulation models can have few or many parameters, and deciding what values those parameters should have can be complicated, or impossible (meaning that we have to guess or try out different values until one seems to work). In $A = \pi r^2$, 2 is a parameter, whose value we happen

---

[19] But this term is often depredated. You may read the expression "assisted migration" in the conservation literature. What they usually mean is assisted dispersal: permanent relocation.

to have known since the Greeks figured it out. A famous relationship in ecology, the species-area curve,[20] is very similar to $A = \pi r^2$, but the parameter has to be identified for each case (species).

Correlation

Correlation refers to a mathematical relationship between two sets of measurements. Typically, each set comprises multiple values of a variable. The correlation is the simplest statistical calculation relating variables. Each measurement on one variable must be associated with a measurement on the other. That pair is known as an "observation", or "record". A positive correlation means that larger values of one variable are associated with larger values of the other, whereas a negative correlation means that larger values of one are associated with smaller values of the other. Correlations are *linear*, so the strength of a correlation indicates how close a plot of the two variables would be to a straight line. Perfect correlations are 1 (positive) and −1 (negative). Correlations are used widely within all the scientific disciplines that I discuss in this book.

Feedback

A feedback exists between two processes or phenomena when each affects the other. In scientific study of feedbacks there is typically a primary process affecting a secondary process, often with implied causation.[21] The secondary process "feeds back" to the primary, creating a *feedback loop* of repeating effects. Feedbacks can be positive, amplifying the primary process, or negative, damping it. For example, in climate-change study, combustion, whether of biomass or fossil fuels, produces both positive and negative feedbacks. CO_2 is released, adding to its atmospheric concentration and amplifying the greenhouse effect (see Chapter 3). At the same time, sulfates and other *aerosols* (suspensions of fine solid particles or liquid droplets) are emitted, blocking incoming solar radiation and damping the greenhouse effect.

---

[20] A graph that shows how the number of species in a specific area changes with the area, i.e., the more land, or water, the more species.

[21] The phrases "correlation is not causation" and the Latin *post hoc ergo propter hoc* (after this, therefore because of this) are ever-present cautions to those making scientific inferences. True feedbacks, by definition, involve at least some causation, or influence, although causes and effects can be many and complex. In our example in the text, a delayed feedback to climate change from biomass burning is the change in reflectance of the land surface (*albedo*) from loss of vegetation, either to reflect more solar radiation (negative feedback) or less (positive feedback).

Gradient

A gradient is a recognizable or identifiable change in a variable, such that we can predict its value somewhere that we haven't observed. For example, there is a latitudinal gradient of average temperature from the tropics to the polar regions. Along this gradient, it is a safe guess that on average, the temperature will be lower as latitude increases (i.e., from 0° to 90° N or S). Gradients interact with other gradients to confuse things; e.g., elevational gradients can act like latitudinal gradients, and maritime-continental gradients affect daily day-night temperature gradients.

Succession

Ecological succession is the change over time in the defining attributes of vegetation over a particular geographic domain. The oft-used example in ecology textbooks is the change over time from a non-forested site to forest, via stages of seedling establishment and growth (early stage), the beginnings of selective mortality of individual trees from competition with others (middle stage), and eventual changes in species composition to species that can grow and survive in a shaded understory, i.e., that are "shade-tolerant" (late stage). Nature is rarely as orderly as textbooks, however, and succession can recycle or turn back on itself in many ways.

Disturbance

We define ecological disturbance, opportunistically for this book,[22] as a relatively discrete event in space or time that changes an ecological state, pattern, or process noticeably. A disturbance can be an integral part of ecosystem dynamics, such as periodic wildfires on a fire-prone landscape, or largely external, such as a hurricane making landfall. Disturbances often reset succession to an earlier stage.

Treeline

Treeline is the boundary between areas that support trees and those that do not. In mountains there are often both upper and lower treelines. Above the upper treeline, climate is too cold (actually too harsh in ways that include cold as a factor, e.g., too

---

[22] There are probably about as many definitions of disturbance as there have been definers. This relieves us of any need to pay homage to a particular one. Instead, we choose one that helps us focus on its relationship to climate change.

wet, too dry, too windy, too snowy), and below lower treeline climate is too hot (similarly, too hot and dry—usually not too wet—or with soils that cannot retain sufficient moisture). Broad constraints on treelines are elevation (both upper and lower) and latitude (northern or southern equivalents of upper elevational treeline).[23] Locally, treelines are quite complex, a subject of much current research to this day, and far from linear or precisely defined.

Rain Shadow

A rain shadow refers to an area on the lee side of an obstacle to the passage of rain, just as a light shadow has its access to the sun or another source of light blocked. In the West, where westerly winds[24] dominate, rain shadows are normally on the east sides of mountains. Rain shadows' locations can therefore be predicted from knowledge of atmospheric circulation. Their strength is a function of the moisture in the air, the local topography, and the wind speed. If major atmospheric circulation changes with a warming climate, rain shadows will move.

Climatic Envelope

This somewhat awkward term is now common parlance for the ranges of climatic variables favored by plant or animal species. For example, one two-dimensional climatic envelope would be 10–50 mm per year of precipitation and mean annual temperature of 35–45 °F. Real climatic envelopes are assumed to be *high-dimensional* (many variables), and are usually approximated in research studies by fewer but key variables, chosen by statistical methods.

Connectivity

This is a basically a measure of how easy it is to get from point A to point B. For example, an underground light-rail network is highly connected when all the trains are running; the city streets above it are less connected if crowded with traffic. So connectivity in space can change due to factors that operate at different time scales. We use this concept to analyze the movement of *vagile* organisms (those that can move: most animals and no plants) across landscapes.

---

[23] Upper latitudinal treeline is evident in the Northern Hemisphere; in the Southern Hemisphere there are no land masses where it would fall, in the Southern Ocean.

[24] Just to be clear, "westerly" winds are winds out of the West, or "west winds". So they are not "blowing in a westerly direction" (an awkward phrase all too often encountered).

Limiting Factor

Limiting factors are constraints on some ecological process. Each process can have one or more[25]; taking away the strongest limit will often reveal the second-strongest, etc. For example, tree growth is typically considered to be *water-limited*, *energy-limited*, or some combination of these. Hot dry sites would be water-limited whereas cold shaded sites would be energy-limited. Some tree species are water-limited in parts of their ranges, or during certain seasons, and energy-limited otherwise. Limiting factors do not have to be climatic. They can be (lack of or excess of) nutrients, presence of predators, competition of various kinds, or genetic predispositions, for example.

Scale

Scale is a key concept in almost all of science, and ecological science is no exception. Many entire books have been written on the topic, so all we can do here is to touch on key ideas that illuminate the scale issues discussed in this book. Analysis of scale issues requires the ideas of *grain* (or "grain size", the smallest unit of space or time being studied) and *extent* (the "domain size", or size of the entire study area). In its typical uses in ecological science,[26] scale refers, sometimes not entirely clearly, to both grain ("fine-scale" versus "coarse-scale") and extent ("small-scale" versus "large-scale"). The jargon "downscaling" refers to making the grain finer for some object of study[27] while reducing the extent. (In models, grain is referred to as "grid spacing", and in observational studies like remote sensing as "resolution"). Other terms we will encounter are *multi-scale*, referring to concurrent or parallel analysis of study objects at different scales, and *cross-scale*, referring to analysis, often mathematical, of how the objects change as the grain and extent of observation change.

---

[25] By definition, no process is without at least one limiting factor. If it were, it would be infinite.

[26] Geographers will notice that "small scale" in ecology means the opposite of what it does in geography, i.e., a larger-scale map covers a smaller extent than a smaller-scale map.

[27] These objects of study can be in the domain of space, or time, or both. An example of the spatial domain is a landscape with different mapped features such as patches of forest, grassland, and rock and ice. In the temporal domain, we study *time series* (repeated measurements, usually at regular intervals [grain] for some period of time [extent]). In both cases, observations closer to each other may be more similar than those further apart. This property is *auto-correlation*.

Uncertainty

Uncertainty has a different meaning in the scientific literature from its use in common parlance, and can be a source of confusion as a result. As opposed to meaning that we don't know whether we are right (common parlance), the scientific use refers to a range of values, typically of variables or parameters, that could be right. A simple example of the latter is the classic bell-shaped curve, defining the *normal distribution* (Fig. 1.2). An estimate of the "right" value is at the top of the curve, but that value is only most likely to be right. Other values along the curve could also be right, and the range of uncertainty therefore bounds the set of plausible values on the left and right.[28] When many values are estimated instead of one, such as in climate models, uncertainty estimates quickly become very complex, and many approximations may be needed. In those cases, qualitative expressions of uncertainty, such as "almost certainly", "very likely", "likely", etc., may be substituted for quantitative estimates, particularly in communicating results to diverse audiences.[29]

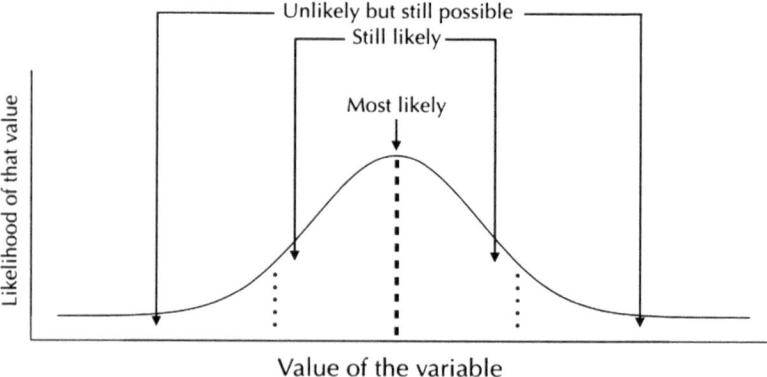

Fig. 1.2 The classic "bell-shaped" curve, or "normal" distribution. When we don't know the exact value of some variable, like temperature, we still may know the distribution of values, and have an equation for it. When the distribution is "normal", we have an estimate of the most likely value, but we concede that it could be different. The farther away we guess from the most likely value, in either direction, the less likely our guess is to be correct

---

[28] In reality, the curve in this example extends infinitely far in both directions, but the probability of being in the "tails" gets smaller and smaller the farther one extends. To adjust for this, uncertainty bounds are often given percentages of the number of possibilities they include, and stated as "confidence intervals". For example, a 99% confidence interval will include (for all practical purposes) all but 1% of the values, whereas the narrower 95% confidence interval will "miss" 5%.

[29] For example, the Intergovernmental Panel on Climate Change, or IPCC, often uses such categories as the primary means to communicate uncertainty, while providing numerical equivalents of the categories for those interested. Links to all the IPCC reports can be found on their website: http://www.ipcc.ch/.

Stationarity

This somewhat technical term refers to the property of things that don't "move". In common parlance it is usually an object in space that is "stationary" rather than in motion. The scientific use refers to processes that can be defined mathematically such that some *variables* depend on others in ways that do not change. A simple example of a stationary process is radioactive decay. Although one cannot predict the exact moment that an atom of the material will emit radiation and decay to an atom of a different element, the average rate of decay is known and does not change (a process whose average doesn't change has "first-order" stationarity). In the Earth sciences nothing is so simple. In our mathematical models it is often very convenient to assume stationarity, but it can be difficult to prove. In a changing climate, empirical observation shows us that many processes for which stationarity would also be quite convenient are evidently not stationary. This is bad news for both the simplicity and reliability of our models. For example, we see frequently in the mass media that there will be more wildfires everywhere in a hotter and drier climate. This seemingly intuitive assumption is borne out in statistical models that predict wildfire extent from weather and climate, but those models vary hugely in their strength, and in their *parameters*, depending on how hot and dry it was where their data were collected.[30]

Detection and Attribution

These are two pillars of science that I will refer to often. They are two separate problems that can be confounded. In the normal order of things,[31] detection comes first: something is observed. Then we ask why it occurred (attribution). A frequent problem with detection is the relative strength of *signal* (what you are trying to observe) to *noise* (what is interfering). For example, with enough static or feedback in a sound system, the words of a speaker may be unintelligible.[32] A key problem

---

[30] Full disclosure: colleagues and I have published research in this area, but it is not accepted by everyone (yet). For example, see D. McKenzie and J.S. Littell. 2017. "Climate change and the eco-hydrology of fire: will area burned increase in a warming western USA?" *Ecological Applications* 27:26–36. This is a scientific re-statement of the reason that Death Valley, CA, USA, is not constantly on fire from being so hot and dry.

[31] But not always. As an example (OK a bit afield), researchers knew that gravitational waves "should" exist. They were predicted by one of the most solid theories of all time: General Relativity. So the *attribution* was easy; they came from interacting massive objects. But their *detection* did not occur until 2018, 100 years after the theory that "attributed" them, because they are so faint in the surrounding noise.

[32] Note from this example that by increasing the signal you may not increase the *signal-to-noise ratio*, which is key. Your speaker may raise his or her voice thereby creating disproportionately more feedback in the system. This issue arises often in the Earth sciences.

with attribution is that "correlation is not causation". As I said above, we use correlation all the time in the Earth sciences, but if we attribute one of two correlated sets of observations to the other, we do so imprudently unless we have other factors to consider. For example, your puppy's growth will be correlated with its caloric intake. We have good reason to attribute that to growth physiology. But your neighbor's kitten's growth is also correlated with your puppy's growth. Does one cause the other? We will see examples below, particularly in Chapters 8 and 9, of how disentangling detection and attribution can be a challenge.

Chapter 2
The Mountains

*"Sometimes I do get to places just when God's ready
to have somebody click the shutter."*
—Ansel Adams

The Western Hemisphere differs from the Eastern Hemisphere in that it is divided by a mountain chain that runs the length of the continent from northern to south. Through much of its length, the *American Cordillera* separates fairly narrow coastal provinces from broader ones to the East, creating sharp divisions in climate between *maritime* (west of the Continental Divide) and *continental* (east of the Continental Divide).[1] Within the conterminous United States (CONUS), however, the width of the Cordillera expands to nearly 1000 miles, across which there are three distinct north-south sub-chains, or "belts". These belts continue north through Canada and Alaska, but it is in the U.S. where they are clearly separated by wide inter-mountain regions. In turn, each of these belts comprises individual mountain ranges, generating a topographic landscape of immense complexity across the western U.S.

The *Laramide Belt* in the U.S. includes the Rocky Mountains, sometimes called the "American Rockies", which we divide in this book into northern and southern portions. The highest peak is Mt. Elbert, 14,439 ft.; there are 53 peaks over 14,000 ft. in the American Rockies, all in Colorado. The *Nevadan Belt* in the U.S. comprises the Cascade Range in Washington, Oregon, and northern California, and the Sierra Nevada in California. The highest peak is Mt. Whitney, 14,505 ft. There are 12 peaks over 14,000 in California in the Sierra Nevada, and one (Mt. Rainier, 14,411 ft.) in the Cascade Range, Washington. The *Pacific Coast Ranges* run parallel to the Pacific coast, are not particularly high (all under 8000 ft.), and include both the wettest region in the western U.S.—the Olympic Peninsula, Washington—and dry non-forested coastal mountains in Southern California.

---

[1] I am greatly indebted to the many developers and authors of Wikipedia for general information in this chapter, such as the heights of peaks, names and numbers of mountain ranges, and geologic time scales of formation. I invite interested readers to delve further by searching Wikipedia for "North American Cordillera", and go from there.

© Springer Nature Switzerland AG 2020
D. McKenzie, *Mountains in the Greenhouse*,
https://doi.org/10.1007/978-3-030-42432-9_2

The large area between the Laramide and Nevadan belts is mostly non-forested. Three prominent geologic regions therein are the *Columbia Plateau* or *Columbia Basin*, the *Colorado Plateau*, and the *Great Basin*. The first two are cut through by the major rivers that give them their names. Though mostly desert, the Great Basin contains many small north-south arid mountain ranges, the highest point of which is White Mountain (14,252 ft.), in the Snake Range, eastern California.

The formation of the Western Mountains is a long history of geologic events, completely unlike the single uplifts that have produced many individual ranges. For example, much of the northern Rocky Mountains dates to the early Paleozoic Era (540–440 million years BP); the Sierra Nevada date from the Mesozoic Era (250–65 million years BP); much of the Cascade Range dates from the Cenozoic (since the extinction of the dinosaurs at the Cretaceous-Paleogene boundary, 65 million years BP). Within each of these areas, however, there is much variation from old to new, at multiple spatial scales. On the surface, slow geological processes would seem to have little to do with rapid climate change, but they have long-term influences, such as determining the texture and depth of soils, which constrain the effects of climate change on vegetation and hydrology.

Climate in the Western Mountains is controlled, at broad scales, by the interaction of global circulation patterns and finer-scale *orographic* effects (e.g., rain shadows). The major *jet streams* of the Earth, at roughly the altitude of the *tropopause*,[2] flow from west to east and are strongest in the temperate latitudes. Westerly winds therefore drive weather in the Western Mountains, with heaviest precipitation falling on their western slopes at middle elevations,[3] and *rain shadows* forming east of the crests of N-S mountain ranges. Fine-scale variability in precipitation is, however, difficult to predict in complex terrain because not all ridges run north-south; different slopes, aspects, and their spatial patterns create complicated paths for the movement of wind and moisture.

As mountains *per se*, the Western Mountains are modest compared to others around the world, or even in the rest of the American Cordillera. The Andes of South America have 72 peaks above 20,000 ft.; the highest (Aconcagua) is 22,841 ft. The Himalaya of central Asia have 26 peaks higher than Aconcagua, with a legendary 14 above 8000 m (~26,000 ft.). The highest, Everest (29,029 ft), is almost exactly twice the height of Mt. Whitney. Nor are the Western Mountains as cold as their global counterparts. For example, Mt. Denali, in the Alaska Range, has recorded −75.5 °F, with wind chill down to −118.1 °F, at a weather station at 18,733 ft. Mt. Vinson

---

[2] The tropopause is the boundary between the troposphere (lower atmosphere) and the stratosphere (upper atmosphere). Its altitude varies from about 52,000 ft. at the Equator to about 30,000 ft. at the poles. Jet-stream winds, often above 150 mph, are typically felt near the summits of the highest Himalayan peaks (28,000+ ft.), making climbing impossible.

[3] The concentration of heaviest rain at middle elevations reflects two factors associated with increasing elevation: the decreasing moisture-holding capacity of the atmosphere and sparser obstacles to air movement (landforms). This has been observed most cleanly on volcanoes like Mt. Rainier, with its roughly conic shape allowing more and more air to pass unobstructed with higher altitude.

(16,049 ft) in Antarctica is likely much colder; the coldest temperature on Earth (−128.6 °F) was measured at Vostok Station, Antarctica, which is at 11,444 ft.

For our story in this book, the Western Mountains live in a "sweet spot" between landscapes fully domesticated by human society and those remote areas ("all rock and ice and storm and abyss"[4]) that are inaccessible to most queries about climate change and related topics. The Western Mountains still have expanses of uninhabited areas over which we have ample opportunity to observe climate and ecological processes across a variety of scales. Much of what we know about the effects of climate change therein comes from the convenient "natural experiments" ongoing across mountains of the West.

The Cascade Range: America's Alps

The crest of the Cascade Range in the U.S. runs almost due north-south from the Canadian border in northwestern Washington to northern California. The crest is dominated by volcanoes, from Mt. Baker (10,781 ft.) in the north to Lassen Peak (10,457 ft.) in the south, with 15 volcanoes higher than 10,000 ft. elevation. These volcanoes form the U.S. segment of the "Ring of Fire,"[5] which forms a horseshoe shape around the Pacific Ocean (with the gap in the Southern Ocean). No non-volcanic peaks are above 10,000 ft., but 19 more[6] are above 9000 ft. Peak elevations are therefore mostly lower than those in other mountain ranges of the West (Fig. 2.1).

Topography is very complex in the northern part of the range, with the larger north-south orientation of the crest crossed by many subsidary ridgelines at all angles to the crest (Fig. 2.2). Moving south into Oregon across the Columbia River, which runs through the only gap in the range even close to sea level, the spatial pattern of ridgelines become simpler, with the continuous crest ending about 60 miles north of the California border. Only the Mt. Shasta (14,051 ft.) and Lassen Peak massifs lie south of there. The mountain range is compact and continuous, such that east-west travel from lowland to lowland, either by foot or vehicle, involves continuous ascent followed by continuous descent. There are few roads across the crest, and large protected areas (wilderness or national park), especially in the north.

---

[4] The full quote is "K2: just the bare bones of a name, all rock and ice and storm and abyss. It makes no attempt to sound human. It is atoms and stars. It has the nakedness of the world before the first man—or of the cindered planet after the last", about K2, the Earth's second highest peak (28,251 ft.), in the Karakoram Himalaya, from Secret Tibet, by Fosco Maraini.

[5] The Pacific Ring of Fire has 452 volcanoes in all, 75% of the world total, and about 80% of the world's largest earthquakes occur along it. Plate tectonics, driving the collision of lithospheric plates, are the cause. The Cascade Volcanic arc, including all U.S. volcanoes and three in southwestern British Columbia, was formed by collision of two plates in the Cascadia subduction zone, which is also a potential source of large-magnitude earthquakes in the Pacific Northwest, U.S.

[6] These numbers are from https://www.peakbagger.com/, and are based on a criterion that a peak is considered separate if it has at least 400 ft. of "clean prominence", i.e., stands out from a ridge or massif sufficiently.

Fig. 2.1 Mt. Baker, taken from the north, the northernmost "Rim of Fire" volcano in the Western Mountains, surrounded by dense forest. Photo by the author

The northern Cascade Range, along with the isolated Olympic Mountains, is the most heavily glaciated area of the Western Mountains,[7] due not only to the unparalleled supply of moisture from the Pacific Ocean, but also to the large east-west extent of high-elevation rugged topography in this part of the range. Glaciers extend further south, to Lassen Peak, but they are fewer and their pattern becomes more linear, following the main Cascade crest.

Climate in the Cascade Range is nearly a microcosm of the larger pattern for the Western Mountains, albeit skewed toward the maritime[8] end of the gradient. The north-south orientation of the range brings a wet climate to the western slopes and a partial rain shadow to the eastern slopes. Because of the overall maritime climate, however, climate on the eastern slopes is considered *mesic*: moderately dry, but not approaching that of a desert or arid grassland. Of special import for many aspects of

[7]A scalable map and other tools for displaying glaciers in the Western Mountains can be found at glaciers.geos.pdx.edu, courtesy of WMI scientist Andrew Fountain.

[8]Along a maritime-continental gradient in climate there are characteristic changes in both temperature and precipitation. In a continental climate, temperature has greater extremes, on both diurnal and annual time scales: days are warmer, nights are colder; summers are warmer, winters are colder. In a maritime climate, everything is wetter: more clouds, more rain, and more humidity, even on sunny days.

Fig. 2.2 Mt. Stuart, taken from South Ingalls Peak, in the eastern Cascade Range in central Washington. Photo by the author

Cascade Range ecology, and for climate change, is that summers are often dry,[9] a pattern typical of Mediterranean climates further south, but with a different cause. In a typical summer in the Pacific Northwest, the Pacific Jet Stream migrates north across British Columbia, cutting short the regular supply of wet weather coming off the Pacific Ocean.

The world record for annual snowfall (1140 in., or 95 ft.) is held by Mt. Baker (measured at the ski area), in the winter of 1998–1999. Snow in the Cascade Range is as wet or wetter than elsewhere in the Western Mountains, meaning that the *snow-water equivalent* (SWE, often pronounced "swee") is higher for the same inches of snowfall than it is elsewhere. SWE is a particularly useful *metric* (a number used as a measure of something, usually with *units*, such as in. or °F) for estimating hydro-logical inputs to ecosystems at lower elevations. Freezing levels vary widely year-round, responding to variation between the presence of continental or maritime air masses. Because of this and its moderate elevations, the Cascade Range is vulner-able to "rain-on-snow" events: heavy rains onto an existing snowpack, typically in early winter when that snowpack is less consolidated than it is later. Instabilities produce avalanches and other mass movements, and damaging floods that affect all

---

[9] Not predictably dry though. Many a prospective Cascade backpacker or cyclist, or Pacific Crest Trail traveler, has encountered up to a week of steady hard rain in August, often the driest month.

downstream areas, including cities in the Puget Sound Basin and Willamette Valley, lowlands west of the mountains.

The maritime climate in the Pacific Northwest produces the most productive forests in the Western Mountains. At low elevations, trees grow to be giants, surpassed in size by only the redwoods of the California coast and the giant sequoias of the Sierra Nevada. Whereas these latter two species are confined to small patches of suitable habitat, the giants of the Pacific Northwest occur on most western slopes at low elevations from central Oregon to northern Washington.[10] Like most forests in the West, those of the Cascade Range are mainly *coniferous*, i.e., made up of cone-bearing needleleaf trees. Worldwide, coniferous forests extend across northern North America, northern Europe, particularly Scandinavia, and western Russia and Siberia, and are the dominant type at most latitudinal treelines. Conifers are hardier than deciduous trees in dry cold climates,[11] which explains their dominance in high northern latitudes, but their overwhelming dominance in the Pacific Northwest, particularly at low elevations, is usually attributed to the relatively mild winters in a maritime climate. This allows photosynthesis almost year-round for species that have leaves, i.e., evergreen conifers, giving them a competitive advantage over deciduous species. With few exceptions, deciduous trees in the Cascade Range are dominant only in frequently disturbed areas such as avalanche chutes and *riparian* (literally "river-side") areas. There are enough of these, however, between the many waterways and frequent winter avalanches, that the proportion of Cascade-Range forest cover that is deciduous species is about 5%.

A dominant impression of "greenness" may be had from the Cascade-Range landscape. Mature and old-growth conifer forests are thick and cover much of the range below treeline. In all but the most recently disturbed sites, whether by logging or natural disturbances such as fire or avalanches, new vegetation has established. At finer scales, bare rock, fallen trees, and uprooted soil are colonized by tree seedlings, shrubs, grasses and herbaceous species, and *non-vascular* plants (lichens and mosses). The latter are particularly evident on what would be bare rock in drier areas of the Western Mountains.

---

[10] The largest trees in the Pacific Northwest are found in the less than 5% of the original forests that remain after more than a century of logging. An inventory of the remaining giants is in *Forest Giants of the Pacific Coast*, by R. Schmid and R. van Pelt, 2002. Also see the website ascendingthegiants.com.

[11] Conifers vary widely in their vulnerability to cold, with those in cold climates having physiological adaptations to cold that have not evolved in species in more moderate climates. In general, though, needles are protected by a waxy coating, or *cuticle*, which shields them somewhat from cold (and also makes them generally less palatable to browsing animals). Some climates are so harsh than even this is not much of an advantage, and deciduous conifers (e.g., larches) are often dominant there.

Most trees in the Western Mountains are water-limited. In the Cascade Range, because of the relatively mild climate and the sheer density of vegetation,[12] the balance tips toward energy limits. The two most thoroughly studied species in this regard are Douglas-fir and mountain hemlock. Douglas-fir is the iconic Pacific Northwest big tree, widespread and dominant in many places except higher elevations; its growth characteristics are of interest to both scientists and timber growers. In its most productive stands at low elevations, it is energy-limited in that competition from other trees, such as its neighbors in plantations for producing timber, is the main constraint on growth. In the rest of its range, the *montane* zones at middle elevations, it appears to be principally water-limited (as it is in most of the West) except in the wettest areas. Mountain hemlock is the characteristic species in cold snowy climates at the highest elevations (treeline and just below it). Like Douglas-fir, it experiences opposite limiting factors in different parts of its climatic envelope. At the snowiest sites it grows better in warmer and drier years, whereas at drier sites, near the southern edge of its range, it grows better in colder wetter years.

In contrast to their high productivity, Cascade-Range forests are not very diverse in the tree layer. Compared to Appalachia, for example, with its 158 tree species, the Cascade Range has 18 species of conifers and 15 species of deciduous trees (other than willows, of which there are 9), some of which are quite rare. Most of its biodiversity is in the understory, making a total of over 1600 vascular plant species, and many more mosses, liverworts, and lichens. Within the diverse understory layer there is a concomitant diversity of *life-history strategies*, or evolved sets of responses and adaptations to the environment. For example, many herbaceous species and shrubs grow best in full sunlight; they colonize or persist in forest openings, whether from natural or human (e.g., logging) disturbance. Other species are adapted to low or very intermittent light ("sunflecks"). These species are most common in undisturbed late-successional forest, in which a multi-layer canopy favors a "grazing" strategy of intermittent light-harvesting. Even these species will not persist in very dense single-aged forests, however, such as are found in dense plantations or other younger stands that have grown back after clearcut logging or high-intensity wildfire.

Human land use in the Cascade Range began when Native Americans settled the Pacific Northwest around 11,000 years BP, and engaged in the region's first "forest management". Settlements were concentrated in the lowlands, on the west side around Puget Sound and on its islands, and on the east side in low-elevation ponderosa-pine forests and grasslands. Fires were set on both sides of the mountains, both to encourage food crops and to improve hunting of game animals that prefer open terrain.

---

[12] A popular measure of the density of forest cover is the "leaf-area index" (LAI), which estimates the average number of vertical layers of leaves in a forest canopy. This is different from canopy cover, though related. For example, a forest stand with 100% canopy cover, with a single horizontal layer of leaves, would have LAI = 1, as would a stand with 50% cover and exactly two layers. Forests in the Cascade Range can have LAI up to 9, the most in the West.

Things changed rapidly with the California Gold Rush in 1848, during which the first sawmills were built and timber was shipped from Puget Sound's harbors. What seemed to be an inexhaustible supply of high-grade timber, first on private lands then expanding to national and state forests, was acknowledged to be limited gradually through the twentieth century.[13] Finally, in 1990, strong restrictions on logging public lands were enacted, ostensibly to protect habitat of the northern spotted owl under the Endangered Species Act. After a century and a half of logging much of the forests in the Pacific Northwest had been cut at least once,[14] however, so even without legal constraints opportunities for high-grade ("old-growth") timber extraction were running out.

Recreational use of the Cascade Range has a lower impact than many other places in the Western Mountains, because much of the terrain, particularly in the northern part, is rugged and roadless. Large areas are accessible only to intrepid foot travelers. This is partly due to the greenness alluded to above. Fallen trees and thickly vegetated avalanche chutes are worse barriers to travel than in most other western mountain ranges. For climbers and backcountry snowsports enthusiasts, there is a lifetime's worth of summits and snowslopes.[15] The most popular backcountry trails are, however, closely regulated to prevent sheer numbers of people, even traveling lightly (i.e., hiking), from degrading sensitive areas, which tend to be the most sought-after destinations.

The Sierra Nevada: The Range of Light

The crest of the Sierra Nevada runs about 375 miles NNW to SSE in central and eastern California. Even more than the Cascade Range, these mountains are compact and continuous, crossed by nothing more than trails between their southern end, roughly at Tehachapi Pass, and Tioga Pass, a seasonal highway that is 10,000 ft. at its summit. About 60% of the range, then, and the area with all its highest peaks, is unbroken by through roads. There are 11 peaks over 14,000 ft. in this southern "core" of the range. Because of the arid climate and southerly latitude, there are only 13 named glaciers in the Sierra Nevada (compare to 165 in the Cascade Range); the largest and southernmost is Palisades Glacier, about 0.4 square miles in size.

---

[13] And coming to a near stop in the 1980s with the conservation movement and associated legislation (see Chapter 9).

[14] The exact amount of remaining old-growth forest in the Pacific Northwest is debated, and estimates vary widely depending on the techniques used to make them. A good analysis of the estimates is in R.A. Norheim, 2001. How institutional cultures affect results: comparing two old-growth forest mapping projects. *Cartographica* 38:35–54.

[15] One legendary climber, Fred Beckey, almost single-handedly catalogued the climbing routes in the Cascade Range. His Cascade Alpine Guide, in three volumes, describes thousands of routes on hundreds of peaks (many of which saw Beckey on a first ascent), from scrambles to hard rock or ice, or both. Backcountry snowsports are equally challenging because of the variable (often bad) weather, complex terrain, and snow conditions that can be difficult to read for avalanche danger.

Fig. 2.3 Lake and skyline near Bishop Pass in the southern Sierra Nevada. Photo by the author

What these mountain have in abundance, though, are *rock glaciers* and related terrain features.[16] In an arid climate, these features can be significant sources of water, both for groundwater storage and for creating wetlands (see Chapter 3) (Fig. 2.3).

Climate in the Sierra Nevada is *Mediterranean*, that is typical of areas around the Mediterranean Sea, with hot summers and relatively mild winters. Most precipitation falls as snow between October and May, although violent thunderstorms can occur in summer.[17] The seasonal patterns of precipitation have different drivers from the Cascade Range. Instead of N-S movement of a Jet Stream *per se*, the dearth of precipitation in summer is caused by movement of a subtropical high-pressure cell,

---

[16] Rock glaciers are what the name implies: moving mixes of rock and ice. They can be either frozen rock debris or ice glaciers covered by *talus* (fields of small-to-large boulders). A field study in 2008 identified 421 rock glaciers and *periglacial rock-ice features* in the Sierra Nevada, at elevations between 7300 and 12,900 ft. C.I. Millar and R.D. Westfall. 2008. Rock glaciers and related periglacial landforms in the Sierra Nevada, CA, USA; inventory, distribution and climatic relationships. *Quaternary International* 188:90–104.

[17] These can be the bane of backpackers and climbers, especially above treeline where there is little protection from lightning. Some notorious lightning traps are the granite domes in Yosemite National Park and elsewhere in the Sierra Nevada, where an unstable atmosphere can bring rapid changes in weather that catch travelers exposed and unprepared.

or *anticyclone*, over the Pacific Ocean west of California, the *North Pacific High*. In winter, this cell moves south, making way for maritime moisture-laden air to move across California. As with other N-S ranges of the Western Mountains, there is a marked rain shadow on the east (lee) side, even in these mountains where the western slope is still dry, especially in summer. Climate can be very different from year to year, particular when there is a strong polarity in the *El Niño Southern Oscillation* (ENSO—see Chapter 3), such that snowpack in the high country can persist year round, or vanish by mid-August.

Topography is complex, and exceptionally rugged in the southern part of the range. Even there, however, the largest features are basically linear, unlike the northern Cascade Range. The eastern escarpment in the southern half (seen from anywhere between the towns of Olancha and Bishop in the Owens Valley) is precipitous (Fig. 2.4), with relief of up to 11,000 ft. in just a few horizontal miles. In some places west of the true crest, there are subsidary N-S crests paralleling the main crest, for example the *Great Western Divide* (sometimes called the *Triple Divide* because it separates the Kaweah, Kern, and Kings Rivers). The highest peak therein, Mt. Kaweah (13,807 ft.), is not much lower than nearby peaks of the main crest. West of the crest, the Sierra Nevada slope down more gently (than on the east) to the Central Valley. In the northern half of the range, landform patterns are less

Fig. 2.4 Mt. Whitney and the eastern Sierra Nevada crest at its highest point, with arid hills of the Owens Valley in the foreground. Photo by the author

linear, and there is a prominent gap at Mammoth Mountain.[18] The last high peaks (above 12,000 ft.) are at the northern end of Yosemite National Park, whose domes and other granite massifs are renowned.

In most Mediterranean climates, summer drought stress is the principal limiting factor for trees and other vegetation. In the Sierra Nevada, the dominant conifer species appear along gradients of water-balance deficit,[19] a metric that estimates how much less water is available to plants than is optimal for growth, depending on their drought tolerance. This tree-replacement pattern is best seen on the steep eastern slopes that rise out of the desert of the Owens Valley. The lowest slopes are dominated by sagebrush, with an occasional Joshua tree. Slightly higher are juniper, pinyon pine, and knobcone pine. Ponderosa pine dominates the middle elevations (~7000 ft.) on the east side, followed by lodgepole pine, and finally, in the subalpine zone above ~9000 ft., whitebark pine and limber pine, which are widespread at high elevations throughout the West, and foxtail pine, which is *endemic* (found nowhere else) to the southern Sierra Nevada and the Klamath Mountains in northwestern California.

On the more extensive and gentler western slopes, water-balance deficit does not track elevation quite so clearly, so that spatial patterns of dominant tree species are more complex (and there are more conifer species altogether). What "elevation bands" exist are still very fuzzy on their edges (true on the east side too but lesser so), and move up and down in response to topography and topographically driven climate.[20] The western foothills are dominated by gray pine, blue oak, and other oaks, willows, and other deciduous species. Higher up (~5000 ft.) where true forests begin, are ponderosa and Jeffrey pine, incense cedar, sugar pine, white fir, Douglas-fir, California black oak, and the endemic giant sequoia. The next elevation zone has red fir, lodgepole pine, the wide-ranging Jeffrey pine and Sierra juniper, mountain hemlock (though not so abundant as in many parts of the Cascade Range), and western white pine. Alpine treelines are often dominated by whitebark pine, as on the east side, but lodgepole pine, whether upright or *krummholz* (a stunted bush-like growth form found at treelines), is also the treeline dominant in many places (e.g. Yosemite National Park).

---

[18] Mammoth Mountain, though not volcanic, is a smaller version of Mt. Rainier in Washington in that it is a moisture trap for westerly winds. It is a haven for snow sports because it receives so much (fairly dry) snow and its northeastern exposures hold the snow often into mid-June.

[19] Water-balance deficit is commonly calculated as the difference between potential evapotranspiration (PET) and actual evapotranspiration (AET). Evapotranspiration is the combination of direct evaporation off the plant surface and transpiration (literally "water-crossing"), the movement of moisture through plant tissues. In a water-limited environment, plants adapt by shutting down or slowing metabolic processes that allow water to "escape" by evapotranspiration, such that AET is less than PET; the hotter and drier it is, the bigger the difference.

[20] *Cold-air drainages* (movement downslope of cold air, often at night) are an important *microclimatic* process that affects vegetation and snow cover directly, and modifies fire spread by changing the flammability of *fuels* (the vegetation, live or dead, that burns). Montane species whose normal elevational range is 1000–2000 ft. higher can be found in cold-air pockets at the lower ends of watersheds affected by them.

The iconic Sierra Nevadan tree species is of course the giant sequoia. It is endemic to these mountains and within them is still found only in *refugia* (literally "refuges" from processes that affect most of the rest of a landscape, whether they are disturbances or just climate that has changed from when a species originally established). These sequoia groves are all in either Yosemite or Sequoia Kings Canyon National Parks, and their preservation in a changing climate is problematic.[21]

As in the Cascade Range, most of the plant diversity of the Sierra Nevada is in the understory. There are 45 tree species—20 conifer and 25 deciduous—but several hundred shrub and herbaceous species (with a particular wealth of sedges). In contrast to the greens of the Cascade Range, the dominant colors in the Sierra Nevada are those of bare rock, bare soil, and sky. Vegetation is generally open and is not itself a barrier to foot travel. It can also be incredibly tenacious: large ponderosa and Jeffrey pines appear to grow from small cracks in granite, thick-barked trees resist fire and others resprout rapidly after fire, and foxtail pines grow (slowly) for centuries to millenia on ridges at treeline, surviving gale-force winds and extreme cold.

The earliest Native American presence in the Sierra Nevada appears to have been around 3000 years BP in the north-central area (north of the highest peaks), much later than settlement of the Pacific Northwest. Northern Paiute tribes lived on the east side, with the Sierra Miwok on the west. Intertribal trade routes passed over the crest, probably mostly in the north, where elevations of passes are moderate, but northern Paiute people crossed from the east side into what is now Yosemite National Park at some point. There were violent confrontations with the resident Miwok people, particularly in Yosemite Valley, certainly a prize piece of territory.

The Gold Rush of 1848 was concentrated in the western foothills. It began with simple panning by individuals, but after most of the accessible gold had been extracted, hydraulic mining began, and continued through most of the nineteenth century, with its associated environmental damage: exposed virtually sterile earth and heavy-metal pollution of waterways. Meanwhile, the 300,000 or so gold-seekers who had arrived in the first decade of the Gold Rush had extinguished the traditional lifestyles of the Miwok and other western Sierra tribes, by occupying their hunting, gathering, and fishing sites.

Exploration of the Sierra Nevada high country began in earnest in the 1860s, and the last and probably most difficult 14,000-ft. peak was climbed in 1931.[22] It was

---

[21] Even without accounting for climate change, giant sequoia preservation is a delicate proposition. A major issue is that although the species is fire-tolerant, high-intensity wildfires such as have occurred early in the Twenty-first century can kill even mature trees. But total fire protection is not the answer either, because the giant-sequoia ecosystem is also dependent on low-intensity fires that thin the understory vegetation, particularly white fir, so that it does not become a *ladder fuel* (literally a step-ladder for flames to reach the canopy) that makes a fire intense enough to kill the sequoias.

[22] John Muir, Clarence King, and George Brewer are the best known early explorers of the Sierra Nevada high country. Norman Clyde is the best known from the early twentieth century, and is an analog to Fred Beckey (see Note 15) in the Cascade Range, though born 38 years earlier. Clyde has over 130 first ascents to his name, most of them in the Sierra Nevada.

only 40 years later that the first wilderness permits were issued for the most popular trailheads on the east side, such as the Mt. Whitney "regular" trail. A combination of an increasing number of devotees to this unique backcountry and the fragility of subalpine and alpine ecosystems to human-caused degradation has created a huge gap between supply and demand for recreationists. Lotteries are now in place, with less than 25% success rates, for the most popular trails. In areas accessible to vehicles, internet reservations for desirable campsites sometimes close (i.e., are all given out) within minutes, if not seconds, of their opening. Heavy (and necessary) regulation of recreational use is thus almost ubiquitous in the Sierra Nevada, whereas in many other parts of the Western Mountains it is still limited to the most popular or fragile areas.

The Rocky Mountains: The Continental Divide

Unlike the Cascade Range and Sierra Nevada, the American Rockies are an array of separate uplifts, spanning six states: Montana, Wyoming, Colorado, New Mexico, Utah, and Idaho. These separate ranges within the American Rockies replicate the continuous structure of the Cascade Range and Sierra Nevada, i.e., mostly N-S orientation and few major breaks. Best known of these are the mountains of Glacier National Park (north) and the Beartooth Mountains (south) in Montana, the Tetons and Wind River Range in Wyoming, and the Sangre de Cristo, College, and Front Ranges and San Juan Mountains in Colorado. Between southern Montana and southern Colorado are all the peaks above 12,000 ft. Between western Wyoming and northwestern Colorado there is a substantial break, known as the "South Pass" (actually two "passes" at 7412 and 7550 ft.), which has high-prairie vegetation and from which only the southern end of the Wind River mountains is visible. Elsewhere between subranges of the American Rockies, there are also wide valleys running north-south, which are below treeline and so have desert or prairie vegetation (Fig. 2.5).

Considering the American Rockies illustrates the importance of *scale*. As a whole, the mountains act as a major force on global atmospheric circulation, creating a 7000–14,000 ft. barrier to westerly winds and precipitation. For understanding mountain ecosystems, however, we are best served by looking at the scales of the subranges, which we do in most of what follows, without ignoring how the distances between them affect how they influence each other.[23] For example, the Wind River Range in Wyoming is like a small version of the Cascade Range or Sierra Nevada, in that it is compact and continuous, traversible only by hikers and climbers

---

[23] *Fragmentation*, and its opposite, *connectivity*, affect how organisms can move between locations and how physical processes such as snowmelt and streamflow may influence more distant locations. For example, wide valleys may be barriers to wind-dispersed seeds (especially upwind), but may not stop large animals such as grizzly bears, cougars, or lynx from dispersing between forest habitat patches unless the valleys are dense with roads or other built structures.

Fig. 2.5 The Bob Marshall wilderness in the Northern Rocky Mountains in southwestern Montana. Photo by Alina Cansler

(Fig. 2.6). It is within the Greater Yellowstone ecosystem, however, thus part of a large tract of national park (Grand Teton and Yellowstone), designated wilderness, and other national forest that is not heavily managed. In contrast, compact sub-ranges in Colorado with many of the highest peaks in the American Rockies—the Front Range, Sangre de Cristo Mountains, and the College and Sawatch Ranges—are relatively isolated from other wildlands, separated by wide valleys or densely populated areas, or both.

There is no obvious division of the American Rockies into *physiographic regions* (having a distinct landscape character, or *geomorphology*). Some authors recognize three: northern, central, and southern. In this book we collapse these to two, with the dividing line running E-W through the wide non-forested zone that includes the South Pass. This is a rough boundary between two climatic regimes. In the north, winter storm tracks from the Pacific Ocean reach the Northern Rockies almost full force, having dropped what amounts to a "surplus" on the Cascade Range *en route*. In the south, much winter moisture (but less to begin with) is lost over the Sierra Nevada, which are twice the elevation of the Cascade Range on average, and over the wide expanse of the Great Basin. In summer, moist Pacific air continues to reach the Northern Rockies, but the south is more affected by dry continental air masses from the north and monsoonal flows from the tropics. Interactions of these flows with the eastern Front Ranges can produce spectacular summer thunderstorms.

Fig. 2.6 Ellingwood Peak in the Wind River Range. Photo by John Angulo

Climate in the American Rockies as a whole is as variable as the terrain, but one common thread throughout the range is the importance of *teleconnections*, or the correlations between climate variations and anomalies across large distances, sometimes many hundreds of miles. There are two main sources of teleconnections in the American Rockies. In the north, climate variability arises from changes in north Pacific surface temperatures and their effects on atmospheric circulation, associated with a low-pressure center of the Aleutian Islands (the "Aleutian Low"). In the south, the strongest drivers are subtropical and tropical surface temperatures, as reflected in the *El Niño Southern Oscillation* (ENSO).[24]

Because the American Rockies cover such a large area, comparing twentieth-century changes in their climate to those of the CONUS overall is useful for revealing how mountains are different from flat terrain, without having a "scale" problem. As with the CONUS, climate in the American Rockies varied through the century

---

[24] I am glossing over a lot of detail here to accentuate the contrast between north and south. ENSO and the north Pacific surface dynamics (linked to another periodic phenomenon called the Pacific Decadal Oscillation) are themselves a teleconnection, although the cause-and-effect relationships are not fully established. To the extent that it affects the eastern Pacific Ocean, ENSO can have a strong influence on the climate of the Sierra Nevada and Cascade Range, and on the other Pacific Coast Ranges. Through ENSO's effects on continental-scale westerly flows, there is a "dipole" effect on the West: in the north the El Niño phase of ENSO is associated with warm and dry weather, whereas in the south it is cool and wet. The reverse happens with the La Niña phase.

until the 1970s, and has been warming steadily since then.[25] Since the 1930s diurnal temperature ranges have decreased, and at high elevations warming has not been consistent with lower elevations or the CONUS since the 1970s, because of several cold years in the early 1980s and then more rapid warming since the 1990s.[26] If global warming is magnified at higher elevations, and evidence is accumulating that it will be,[27] large areas of the American Rockies (mainly in the southern part) will feel its effects more strongly than the rest of the Western Mountains.

The latitudinal gradient of climate from north to south in the American Rockies is reflected in the decreasing numbers of glaciers, despite the increasing elevations of the peaks. There are 60 named glaciers in Montana, with 35 in Glacier National Park (down from 150 in 1850), all below 9000 ft. Wyoming has 37 named glaciers, all above 9800 ft.; Colorado has 14, all in Rocky Mountain National Park and above 10,800 ft.; New Mexico and Utah have none.[28]

The geology of the American Rockies dates back 1.7 billion years, to the formation of the metamorphic rock that extends across North America, and is as diverse as the aboveground landscapes. Beginning about 350 million years ago, various plate-tectonic events (*subductions*, literally "leading under") lifted the subranges that we know today. The broadest uplift probably occurred when an oceanic plate collided at an unusually shallow angle with the North American (continental) plate, causing the main uplift of the American Rockies to be many hundreds of miles inland. This occurred sometime during the Jurassic Age.

There is complex rugged terrain in the American Rockies, to be sure (e.g., the Wind River Range and the Tetons), but there are also large areas of gentler topographic facets (e.g., the Yellowstone Plateau, which is indeed nearly flat) distributed widely along the full length of the range. Many of these gentler slopes are covered by forests that have large stands of a single tree species. In Yellowstone, for example, this is lodgepole pine. At middle elevations throughout the southern part of the range, these can be quaking aspen, the one common deciduous species in a region dominated by conifers. These *monospecific* stands influence the extent and severity

---

[25] T.F. Kittel *et al.*, 2002. Climates of the Rocky Mountains: historical and future patterns, Chapter 2 in *Rocky Mountain Futures*, edited by J.S. Baron, ran an in-depth quantitative analysis of twentieth-century climate in the Rocky Mountains. I draw on their analysis for the summary here.

[26] There is some debate over whether there has been a systematic bias in recent measurements of high-elevation temperature, from the "SNOTEL" sites. Temperature sensors were switched at over 700 sites starting in the mid-1990s, and their calibration is in question. See J.W. Oyler *et al.*, 2015. Artificial amplification of warming trends across the mountains of the western United States. *Geophysical Research Letters* 42:153–161.

[27] A working group for the Mountain Research Initiative (http://mri.scnatweb.ch/en/) identified five factors that could accelerate "elevation-dependent warming": snow-albedo and surface feedbacks, water vapor changes and latent heat release, changes in surface water vapor and radiative flux, surface heat loss, and aerosols. (in *Nature Climate Change* 5:424–430).

[28] The named glaciers are a subset of "snow and ice bodies", which may be just small patches of permanent snow that are indistinguishable from true glaciers when remotely sensed. In Colorado, there are many of these latter outside of Rocky Mountain National Park, but the lowest is still above 10,600 ft.

Fig. 2.7 The upper part of the Loch Vale watershed in Rocky Mountain National Park, and the site of a long-term ecological research program. Photo by Michael Bergman

of ecological disturbances such as wildfires, driven by differences in density and flammability, and insect outbreaks, driven by species-specific susceptibility.

Like the eastern slope of the Sierra Nevada, many subranges in the American Rockies have distinctive lower and upper treelines. For example, the Front Range of Colorado rises abruptly out of the heavily populated plains, but forests do not begin until after 1000 ft. or so of elevation gain, and these are dry ponderosa pine stands that are not dense.[29] Upper treeline in the American Rockies ranges from about 12,000 ft. in northern New Mexico to about 6000 ft. in Glacier National Park, although as we noted earlier, treeline elevations are controlled by many factors at multiple scales, and may differ substantially at the same latitude with different aspects, slopes, microclimate, and disturbance history. Throughout the range, then, there are 2000–3000 ft. of terrain above treeline on the highest peaks. This means that the American Rockies have far more acreage in alpine vegetation—grasses and flowering herbaceous plants—than the rest of the Western Mountains combined (Fig. 2.7).

The American Rockies were home to Native Americans beginning soon after they dispersed across the Bering land bridge (ca. 14,000 years BP).[30] Beginning in

---

[29] The intense Waldo Canyon wildfire that burned 346 homes near Colorado Springs in 2012 was fueled by non-forest scrub vegetation, more than by these ponderosa pine stands. Not so the Black Forest Fire a year later, which burned 486 homes, all in an isolated patch of unusually dense ponderosa pine trees, east of the I-25 corridor but 1000 ft. higher.

[30] Until the 1990s, it was generally thought that the first migration to the Western Hemisphere was about 13,500 years BP, based on spear points at an archaeological site at Clovis, New Mexico. Since then, different theories have been proposed, including some based on genetic analyses that suggest that the first migration was much earlier (ca. 30,000 year BP). As I write, the issue appears to be still open.

the 1500s, invasions of the Spanish from the south brought profound changes, such as horses, rifles, and new diseases, to the Native cultures. In the 1800s, beginning with fur trappers and traders, white Americans inundated the American Rockies, and by the 1880s the Native populations were largely removed from most of their historical ranges. In the 1840s, the Oregon Trail became a major thoroughfare for settlers headed further west, and gold rushes in Colorado, Idaho, and Montana brought thousands of prospectors. Besides gold, large reserves of other minerals and fossil fuels were found in many places. To this day, abandoned mines, with their tailings and other toxic wastes, are a significant environmental hazard throughout the region.

The American Rockies see heavy recreational use year-round. The continental climate, with its dry powder snow, brings snowsports enthusiasts to the 97 ski resorts and backcountry users to wilderness areas. Summer sees everything from off-road drivers (on the 100 s of miles of 4-wheel-drive roads), to extreme rock climbers, to peak-baggers for whom the "14ers" are an analogous objective to the "8000ers" in the Himalaya.[31]

The Pacific Coast Ranges

The Pacific Coast Ranges in the CONUS comprise 38 named subranges, extending from the Olympic Mountains on the Olympic Peninsula in northwestern Washington State to the Transverse and Peninsular Ranges of Southern California. Across these 15° of latitude there are strong climatic gradients, both in temperature and precipitation, although all the ranges are under the influence of westerly winds off the eastern Pacific Ocean. I shall highlight the three most prominent of the ranges, two at the ends of the climatic gradient and one in the middle.

The Olympic Mountains

The Olympic Peninsula is virtually an island ecosystem surrounded by lowlands and water. The area is large and mountainous enough, however, to support strong biophysical and ecological gradients. On the windward side (Pacific Coast), are rugged beaches and the southernmost part of the temperate rain forest that extends north to the Alaska Peninsula. The Olympic Mountains create such a rain shadow, with the southwesterly winds, that precipitation drops from 140 in. per year in the west-side valleys (and 220 in. per year on Mt. Olympus, the highest point in the

---

[31] Of the 53 14ers in Colorado, two have roads to their summits, 14 have trails, and the rest are no harder than *Class 4*, which is fairly easy climbing but sometimes with significant exposure, where a fall could be fatal. "Bagging" all the 14ers is orders of magnitude easier than climbing all the 8000ers, at which only 43 people have succceeded (as of late 2019).

range) to 16 in. per year in the driest (NE) part of the peninsula.[32] Seasonal climate in the Olympic Mountains follows the same pattern of dry summers as in the Cascade Range to their east.

All of the Olympic Mountains below treeline (5000–6000 ft.) are forested, with tree species changing from (south)west to (north)east in response to the steep climatic gradients. Low-elevation forests on the west side are the most productive in North America, with Sitka spruce, Douglas-fir, western redcedar, and other species reaching record-breaking sizes. The forest floor is an almost continuous green, with herbaceous species, mosses, and lichens covering the ground surface and fallen logs (Fig. 2.8). The central core of the range, above treeline, is heavily glaciated. The eastern slopes and valleys are often rugged, with fast-moving streams and vegetation similar to parts of the Cascade Range, including species typical of the more mesic areas on its east side.

As with the high elevations of many of the Western Mountains, the core of the Olympic Mountains is accessible only to intrepid travelers (i.e., hikers and climbers) because of its ruggedness and year-round snow and ice. This and its geographic isolation have limited the establishment of both plant and animal species that inhabit the Cascade Range, or the Coast Ranges further south, or both.[33] This same isolation left the upper elevations relatively intact such that by the time Olympic National Park was designated in 1938, almost a million acres still "qualified" as national-park-worthy.[34] In contrast, much of the very productive forest of the lowlands on the west side has been subjected to clearcut logging, such that the virtual boundary between protected and exploited areas is visible from satellites.[35] This means that whereas the mountains themselves are ecologically intact, connectivity with other mountain landscapes is minimal.

Siskiyou-Klamath-Trinity Mountains

This range, often abbreviated to the "Klamath Mountains", lies in the middle of the latitudinal transect of the Pacific Coast Ranges, in the southwestern corner of Oregon and northwestern corner of California. With complex rugged topography

---

[32] See http://www.olympicrainshadow.com/olympicrainshadowmap.html for a map display of the effects of the rain shadow created by the Olympic Mountains.

[33] Lack of appreciation for the ecological consequences of this isolation led to the introduction of mountain goats to the Olympic Peninsula in the 1920s. Well-meaning people saw this as "natural" habitat for the goats, as well as the more mundane goal of providing fodder for hunters, but ignored the potential effects of goats on alpine plant communities not adapted to their presence. Goats have also become habituated to humans, and in 2010 a hiker was fatally gored. As I write, an effective and humane way to remove the over 300 goats now in Olympic National Park is still being sought.

[34] And in 1988, the US Congress designated 95% of the Park as wilderness. North Cascades National Park, in the Cascade Range, is also a "wilderness park", with only half the acreage of Olympic National Park in the park proper, but much more connectivity with other wilderness landscapes.

[35] For example, see https://visibleearth.nasa.gov/images/87507/olympic-national-park.

Fig. 2.8 Moss-covered trees in the Hoh rainforest in Olympic National Park. Photo courtesy of the National Park Service

like that of the Sierra Nevada to their southeast,[36] though often *transverse* (running W-E rather than N-S), and subject to multiple climatic influences associated with

---

[36] As recently as 100 million years ago, the Klamath Mountains may have been a continuation of the northern Sierra Nevada, but migrated west, as part of their complex geological history. Regional-scale geological structures match both those of the Sierra Nevada and the Blue Mountains of northeastern Oregon.

Fig. 2.9 Continuous mixed-conifer forest in the Klamath-Siskiyou Range. Photo by Jessica Halofsky

the Jet Stream and ENSO, the Klamath Mountains are a classic case of a complex ecosystem. The highest peaks are only just over 9000 ft. (Mt Eddy, 9025 ft.; Thompson Peak, 9002 ft.), but there is very little flat terrain in the range except in river valleys[37] (Fig. 2.9).

Climate in the Klamath Mountains is as complex as the geology and topography. Though experiencing a rain shadow as a whole on the lee side of southwesterly air flows, there are also monsoon-like influences moving north from the Sacramento Valley, creating patches of higher precipitation in what would otherwise be rain shadow in the eastern part of the range. There is also a fog belt along the coast, meaning that coastal summer days are often cool and wet. In general, rainfall is steady and moderate, instead of coming in deluges. Summer thunderstorms further inland can produce dry lightning, conducive to wildfires, which like other ecological processes, have a complex natural history in the Klamath Mountains.[38]

---

[37] This section draws heavily on J.K. Agee, *Steward's Fork*. 2007, University of California Press. I refer interested readers to that book for details on many aspects of the Klamath Mountains.

[38] Pre-conquest fire regimes in the Klamath Mountains were a classic *mixed-severity* type, in which patches of complete mortality are interspersed with those of low-intensity surface fire and all intensity levels in between. These patches vary widely in area, creating the classic "complex mosaic" of vegetation in the aftermath of wildfire. Fire severity also varies among the many vegetation types, from redwood forest, to sparse woodlands and chaparral, to high-elevation (subalpine) forest.

The diverse vegetation in the Klamath Mountains reflects climate and disturbances, but also the complex geological history. A strong influence has been the ubiquity of *ultramafic-derived* (*igneous* [cooled magma or lava] rocks with low silica and potassium and high magnesium and iron), or *serpentine*, soils. These soils are low in essential nutrients such as nitrogen, potassium, and phosphorus, thereby favoring plant species that might be out-competed by more common plants on more benign substrates. Twenty-nine conifer species are found in the Klamath Mountains, the greatest conifer diversity known, including the world's only population of (Brewer's) weeping spruce and most of the range of Port Orford cedar and Baker cypress. The Klamath Mountains are also the northernmost extent of several conifer species and the southernmost extent of several others. This is an area whose biodiversity is sensitive to climate change.

The Klamath Mountains have a history of exploitative land use. Gold was discovered on the American River in 1848, leading to a rush analogous to those in the Sierra Nevada and Klondike. Early mining methods, and the larger-scale more intensive ones that followed, were modeled on techniques developed for the Sierra Nevada, because the geology of gold-bearing deposits was similar in both ranges. Timber extraction began on the accessible river valleys, then later moved onto steeper slopes, creating the usual problems with erosion and other land degradation. Perhaps the most contentious issue in modern times is water use. For example, intensive agriculture on the Klamath River, upstream of where it cuts through the Klamath Mountains on its path to the Pacific Ocean, competes with maintaining adequate streamflows for ecological processes and fish habitat, in keeping with its Wild and Scenic River designation.

Transverse and Peninsular Ranges

Southern California is known for its megacity and associated urban sprawl, but also has a significant array of mountain ranges, including three peaks over 10,000 ft. (i.e., higher than any Olympics or Klamaths), San Gorgonio (11,503 ft.) and San Jacinto (10,834 ft.), both with prominences over 8000 ft., and Mount San Antonio (Mt Baldy), at 10,068 ft. The many subranges in this area are divided into transverse and *peninsular* (these latter following the coastline and continuing to the tip of Baja California). Geologically separating the transverse from the peninsular ranges is the Hollywood Fault, along the northern edge of the Los Angeles Basin. This is separate from the well known San Andreas Fault, and has not produced a significant earthquake in recorded history (Fig. 2.10).

The Mediterranean climate of the Los Angeles Basin extends to its neighbor mountain ranges, so that at higher elevations climate is similar to the southern Sierra Nevada. From the late twentieth century on, abundant snowfall has been restricted to the El Niño phase of ENSO, during which flooding can occur on slopes denuded by fires. Lower treeline can be as high as 7000 ft., with lower slopes populated by often dense chaparral and other shrubs. These lower slopes are vulnerable to

Fig. 2.10 The Angeles Crest, part of a long transverse range that includes the San Gabriel and San Bernardino Mountains, with the Los Angeles skyline and urban air pollution in the foreground. Photo by Nancy Strick

high-intensity fires driven by the *Santa Ana winds*, dry easterly winds off the deserts coupled with very low humidity. The mature chaparral, in particular, is almost a perfect fuel when dry, with its high density and fire-friendly *packing ratio*[39] (Fig. 2.11).

The Southern-California mountains, mainly their foothills, and residential areas built up into them, form a classic *wildland-urban interface* (WUI), wherein vegetation, typically flammable and often dense, is interspersed with structures. This geography sets up much of the area for a perfect storm of Santa Ana winds, fire starts (almost all fires in Southern California are human-caused), and complicated routes of access to fires by firefighters and escape from them by residents. Because of this vulnerability, essentially all fires are suppressed, even those that are far from structures when ignited, and there are no "natural" fire regimes remaining. Whether that makes a difference to fire severity, i.e., does fire suppression exacerbate fire risk

---

[39] The packing ratio is a metric from fire-behavior analysis that measures the spacing of pieces of fuel within a specific volume. The optimal packing ratio for combustion is a density in between solid fuel (e.g., a big piece of wood) and sparse fuel (e.g., a few twigs separated by large gaps). The packing ratio of typical kindling is close to the optimum, hence its name and use.

Fig. 2.11 Tall dense chaparral in the Santa Monica Mountains, an area subject to Santa Ana winds that drive extreme wildfires. Photo by Deb Rivera

in the WUI, has been debated contentiously in the scientific literature, with no clear winner.

The Basin Ranges

The Great Basin is often equated with the "Intermountain West". That said, there are 131 named subranges and over 200 named watersheds within it. The subranges are oriented primarily N-S (Fig. 1.1), thus perpendicular to the continental westerly air flows. The densest "population" of basin ranges lies between the Sierra Nevada on the west and the Wasatch and Uintah Mountains of Utah on the east (figure). Prominent high points are the aforementioned White Mountain (14,252 ft.), Wheeler Peak (13,065 ft.) in Great Basin National Park, Nevada, the state's highest independent mountain,[40] and Kings Peak (13,534 ft.) in the Uintah Mountains and Utah's highest peak (Fig. 2.12).

---

[40] Technically, the highest peak in Nevada is Boundary Peak (13,147 ft.), but it is part of a massif whose highest point is Montgomery Peak (13,441 ft.) in California.

Fig. 2.12 Wheeler Peak in Great Basin National Park. It is part of the Basin Ranges but is also a sky island, with every vegetation type from desert to alpine represented. Photo by the author

The importance of the Basin Ranges for our story is that like the Sky Islands (next subsection), these are ecosystems of limited area that are spatially isolated.[41] Influx of both plant and animal species is limited, meaning that in a warming climate many species will become less and less fit and thus vulnerable to *extirpation* (local extinction). For example, there are eight plant species endemic to the Great Basin, of which two are found only in Great Basin National Park.[42] The corresponding numbers for animals are nine and seven. The seven endemic animals in the Park are arthropods found only in caves.

Climate is expected to be the principal driver of change in the Basin Ranges, because human populations are small and dispersed, except for in cities (Reno, Las Vegas, Salt Lake City) on flatlands in the Great Basin. The mountain ecosystems act as refugia from the hotter lowlands, but will become less effective as global temperatures rise.

---

[41] One could of course argue that Great Basin National Park is also a sky island. We call it part of the Basin Ranges because of its location—far north of the sky islands of the Southwest—but most statements about the Southwest's sky islands also apply to it.

[42] See the national parks websites http://www.nps.gov/grba/learn/nature/endemic-plants.htm and http://www.nps.gov/grba/learn/nature/endemic-animals.htm for the names of endemic plants and animals, respectively.

Fig. 2.13 A watershed on a lesser sky island (than Wheeler Peak) on the Coronado National Forest in Arizona. Photo by Alina Cansler

The Sky Islands

Sky islands are mountains surrounded by different environments that are hostile to the plants and animals of the mountains. A worldwide phenomenon, their iconic expression in the Western Mountains is in the Madrean Sky Islands of Arizona and New Mexico (and continuing into the Mexican states Sonora and Chihuahua). The U.S. has about 27 (depending on the criteria used) sky islands in the Southwest, mostly in southeastern Arizona, with elevations from 5000 ft. to 11,000 ft.[43] (Fig. 2.13).

The Madrean Sky Islands are a microcosm of plant-community gradients, from desert scrub and grasslands to upper-treeline forests. For example, the Pinaleño Mountains of Arizona have the most distinct plant communities per horizontal distance of any mountain range in North America, from oak woodlands at ~5000 ft. to Engelmann spruce and subalpine fir at ~11,000 ft.[44] The very highest sky islands in

[43] If the San Francisco Peaks in northern Arizona are included, the highest is Humphreys Peak (12,633 ft.). With Great Basin National Park included, it is Wheeler Peak (13,065 ft.). The world's highest sky island (depending once again on criteria used) may be Mt Kilimanjaro (19,341 ft.), with a prominence of 19,308 ft. (basically the entire mountain).

[44] As we have seen, the concept of a plant community can be slippery, and "lumpers" and "splitters" may arrive at very different counts. Nevertheless, several factors contribute to the exceptional diversity across distance in sky-island plant communities: the prominence of the ranges themselves, the accentuated *lapse rates* (changes in climatic variables with elevations) between these particular elevations (5000–11,000 ft.) in the Southwest, and the strong contrast in the effects of aspect (especially SW vs. NE) in these water-limited sites.

the Western Mountains (not Madrean) have bristlecone pine and limber pine at upper treeline (above spruce-fir).

As with the Basin Ranges, the chief threat to endemic species in these ranges is a warming climate, which is expected to move suitable habitats up in elevation, progressively reducing their area toward zero (mountaintops). Besides climate, a unique land-use conflict arises from the dry desert air and the clear night skies: the sky islands in the Southwest are ideal sites for telescopes. Construction of telescopes and their infrastructure takes place on the smallest (and already shrinking) habitats, the tops of the mountains. Many of the sky islands, in the Southwest and elsewhere,[45] are also Native-American sacred sites, further increasing the complexity of decisions about their management.

Western Mountain Vulnerabilities

In the chapters that follow, I delve into the ecological patterns and functions in the Western Mountains that are vulnerable in a rapidly changing climate. In all of the ranges highlighted so far, and in those I have omitted, changes are expected in most of these areas. Below I provide a brief overview of what is to come.

Vegetation

In the Western Mountains, most of the vegetation is forests. The overarching control on the distribution, establishment, growth, and mortality of trees is climate. Climate also affects trees indirectly, through its effects on *natural disturbances* (wildfire, insect outbreaks, lethal pathogens). In a changing climate, forests are vulnerable to stronger limits on growth and survival, e.g., increased water-balance deficit. Mortality can occur from intense or prolonged drought, or both. Increasing disturbance, in extent or severity, or both, can reduce or eliminate forest vegetation.

A warming climate may also change species composition. For example, in a warmer climate, water-balance deficit increases, and dominant species may change to those tolerant of drier conditions. Where trees are already at their limit, at or near lower treeline, shrubs may replace trees as the dominant vegetation, particularly after a disturbance that kills mature trees, which are more resistant to extreme conditions than seedlings.

---

[45] A \$1.5 billion telescope, the "Thirty Meter Telescope", with eight times the optical power of any current optical telescope, is planned for completion in 2022 on the Hawai'ian volcano Mauna Kea. At 13,796 ft. asl (and equal prominence by standard definition), Mauna Kea is the ultimate sky island, which if measured from where it arises from the sea floor is higher than Mt Everest. Opponents argue that the observatory will be built on a cultural heritage site. As I write the conflict is unresolved.

Upper and lower treelines themselves reflect climate limitations, and intuitively these would be expected to move up in elevation in a warming climate. Treelines are complex, however, and other climate-driven changes, such as increased wildfire at upper treeline, may counteract the intuitive trends.

Glaciers, Snowpack, and Hydrology

Glaciers are one of the purest indicators of temperature changes, on time scales from seasonal to millenial. Their *mass balance* (literally, the balance between gain and loss of ice) is a sensitive indicator of whether they are gaining or losing ice. With many, though not all, of the glaciers in the Western Mountains receding steadily, we can capture an almost undiluted picture of the effects of global warming.

Seasonal snowpack is more variable than glacier mass balance, as it depends heavily on inter-annual climate variability and teleconnections with quasi-periodic patterns like ENSO, but both snowpack and glaciers affect the often complex downstream hydrology of the Western Mountains. On average, earlier snowmelt, regardless of the total snowpack, means earlier *peak runoff* (the maximum rate of streamflow in a given year) and less available water (unless it is captured by dams) for vegetation in late summer and autumn.

Biogeochemistry

This complicated term refers literally to the organic and inorganic chemistry of physical and biological processes on landscapes, both terrestrial and aquatic. The biogeochemistry of mountain ecosystems is sensitive to inputs (deposition of nitrogen or other nutrients), rates (photosynthesis, respiration, decomposition, weathering), and residues of chemical reactions (secondary aerosols or other reactants). Important climate-induced changes will be disruption of biochemical cycles, either from disturbance or temperature-related rate changes, or pollution by deposition, erosion, or increased concentrations of residues.

Mountain lakes and streams, in particular, are sensitive to increased inputs or rate changes. High-elevation water bodies often are weakly *buffered* (resilient to drastic changes because of a surplus of some protective element) and subject to degradation from increased inputs of nitrogen or other elements. Biogeochemistry also interacts with disturbances in complex ways, making predictions about future processes difficult.

Wildlife

Following common practice, we equate "wildlife" with vertebrates: mammals, birds, reptiles, amphibians, and fish. Though experiencing their worlds at vastly different spatial scales (e.g., migratory birds or wolves and bears vs. amphibians), and in addition to being vulnerable to direct environmental influences (e.g., stream temperatures warmer than a tolerance limit for a fish species), all species are sensitive to three major climate-driven constraints.

Loss of habitat affects individual survival and population size. Habitat can be a type, density, or structure of vegetation, specific features such as pools (amphibians) or dead trees (cavity-nesting birds), or a transient seasonal feature such as snow cover. Loss of connectivity means that even if habitat does exist, it may be inaccessible or require a major effort to reach it. This can range from inconvenient (thereby reducing overall fitness) for a very mobile species such as wolves, bears, or lynx, to lethal, for amphibians that depend on a narrow range of environments. Loss of a specific resource can range, once again, from inconvenient to lethal. Loss of a preferred prey can reduce fitness, whereas loss of a critical habitat feature (e.g., a secure nesting site) or food source can be lethal.

Wilderness Character

This aesthetic component is one of a multitude of *ecosystem services* (aspects of ecosystems that benefit humans either directly or indirectly) provided by the Western Mountains. We value especially the pristine quality of our national parks and protected areas, and this is vulnerable to a host of influences associated with climate change. For example, at local scales, climate change threatens iconic ecological features such as giant sequoia groves and other old-growth forests with the possibility of increased extent, frequency and intensity of disturbance, as well as movement of the baseline climate outside the range of tolerance of the key species. At regional scales, increased wildfire extent would reduce visibility in scenic areas, sometimes to the point of obscuring key landmarks (Fig. 2.14). West-wide, many valued aesthetic resources may change, for better or worse, despite our best efforts to resist that change.

Fig. 2.14 The eastern Sierra Nevada crest (see Fig. 2.4) obscured by wildfires in the summer of 2008. Photo by Doug McKenzie

Chapter 3
It's Getting Warm Down Here

Climate is what we expect, weather is what we get.
—Mark Twain

What we notice about the weather is our most immediate connection to how the Earth's climate may be changing. It is easier for us to notice that things are changing than that they are not, and easier to notice large or abrupt changes than small or gradual ones. The best scientific evidence we have of the particulars of current climate change is, however, of the global and gradual type. Consequently, it can be difficult to reconcile everyday perceptions of climate (weather) with what is compelling evidence that climate is warming, quickly, globally, persistently,[1] and almost certainly as a result of human actions.

The two billion-year-plus history of life on Earth records how species evolved and went extinct in a context of changing atmospheric chemistry, gradually shifting land masses as a result of plate tectonics, and changes in climate. The Earth's climate is a *nonlinear dynamical system,* a complex array of drivers and feedbacks in which the magnitude of a response can be way out of proportion to its cause.[2] It is a truism that "climate is always changing", although there have been remarkably stable periods in the past,[3] but the science of climate change is engaged in the particulars of that change. What direction (warmer or colder, wetter or drier), how fast, how global, how persistent, and how certain are the changes?

---

[1] Meaning that global warming will persist in the Earth system for hundreds of years, even were we to stop raising the CO_2 concentrations in the atmosphere, because of the *inertia* in the dynamics: CO_2 persists in the atmosphere and its greenhouse effect is cumulative.

[2] In particular, climate is known to be a *chaotic* dynamical system: the future state depends sensitively on previous states in a way that makes forecasts uncertain for short-term weather beyond a few days. Longer-term projections, such as those from global climate models, are also affected. Because of this, "official" projections of future warming are usually on the conservative side.

[3] One of which was the period (late Holocene) during which human civilizations developed, roughly 7000 years BP to sometime in the twentieth century (the time at which current anthropogenic global warming could be said to have begun).

© Springer Nature Switzerland AG 2020
D. McKenzie, *Mountains in the Greenhouse,*
https://doi.org/10.1007/978-3-030-42432-9_3

What Do We Know About Climate Change?

The baseline fact about climate change on the Earth is that global temperatures track levels of CO_2 in the atmosphere. As we will see below, this "tracking" is most evident at coarse time scales: centuries and greater. Much challenging and intricate science has been necessary to confirm that this intuitive relationship between temperature and CO_2, associated with the *greenhouse effect*,[4] holds as far back in planetary history as there are good data to test it.

The global and gradual changes mentioned above as evidence for climate change are reflected in the Earth's *energy balance*. Just as we gain weight by consuming more calories than we burn, the Earth gains energy (heat) by absorbing more radiation than it emits. The greenhouse effect is the manifestation of an imbalance in the Earth's energy budget; with high enough concentrations of *greenhouse gases*[5] in the atmosphere, too much[6] outgoing radiation is prevented from escaping into space. In the language of the physical sciences, the greenhouse effect is called a climate *forcing*, or a process that drives climate dynamics directly. The other important climate forcing is from *aerosols*, mixtures of insoluble particles or droplets in a gas (in this case the atmosphere). Aerosols include volcanic ash and sulphuric acid, dust, and many different human-made pollutants such as sulfate aerosols from fossil-fuel combustion.

Greenhouse forcing is fairly straightforward, and by definition always positive, i.e., more greenhouse gases = more forcing = a warmer Earth. In contrast, aerosol forcing can be *positive* (meaning warming, not "better") or *negative* (cooling, likewise). Heat-trapping aerosols, mimicking greenhouse gases, warm the Earth,[7] whereas aerosols that reflect incoming (short wavelength) radiation cool the Earth. Global aerosol forcing is poorly documented, but is generally assumed to be negative overall and only about a third as large as CO_2 forcing. Estimates vary widely,

---

[4] Technically, this mechanism is imprecisely or even falsely named. Greenhouses depend more on limiting convective cooling that on trapping of outgoing longwave radiation, which is how the Earth is warmed. Nevertheless, I will use the term, as others in climate-change science do.

[5] Greenhouse gases are compounds that absorb radiation of the longer (infrared) wavelengths emitted by the Earth. The most important of these for global climate change is CO_2, but it is not the most potent per unit mass. Methane (CH_4—four hydrogen atoms bonded to a carbon atom) is 84 times as potent as a greenhouse gas, though much shorter-lived in the atmosphere than CO_2, so that after 100 years, for example, its effects are "only" 28 times as strong. Other human-made compounds, such as fluorides of carbon and nitrogen, are even more potent.

[6] "Too much" is of course a value judgment. With no greenhouse gases, the average temperature of the Earth's surface would be about 0 °F, about 60 °F below the current average. Detailed calculations of the Earth's energy imbalance suggest that about 350 parts per million (ppm) of CO_2 as an average annual concentration would restore the energy balance and stabilize the Earth's average temperature. See J. Hansen, *et al.*, 2011. Earth's energy imbalance and implications. *Atmospheric Chemistry and Physics* 11:13421–13449.

[7] An underappreciated fact is that the warmth we actually feel is from long-wavelength radiation from the Earth's surface, not directly from the sun's radiation. Cooler temperatures at higher elevations are a consequence of this.

however, with major implications for projecting climate change, as we shall see below.

In a system as complex as the Earth, these two forcings (greenhouse and aerosol) generate many feedbacks. Four are especially relevant to our study of the Western Mountains: *albedo*, biomass burning, *CO_2 fertilization*, and water vapor. Whether they are positive or negative can make a big difference. *Surface albedo* is the reflectivity of the Earth's surface; higher values mean that more radiation is reflected (cooling) and less absorbed (warming). Snow has higher albedo than other land cover, and sea ice has higher albedo than seawater, so reduced snow and ice on a warming Earth are a positive feedback to climate change. *Biomass burning* releases CO_2 into the atmosphere, thereby increasing greenhouse forcing, whose warmer temperatures are generally expected to increase the likelihood and extent of wildfires.[8] This is therefore another positive feedback to climate change. *CO_2 fertilization* is the enhanced productivity of vegetation that is expected if CO_2 is a *limiting factor* for growth and will be more concentrated (i.e., less limiting) in the future. If global biomass increases from this fertilization, then more carbon is sequestered in vegetation, thereby removing it from the atmosphere and creating a negative feedback to climate change.[9] Water vapor is a potent greenhouse gas. Although individual molecules are very short-lived in the atmosphere, a persistent state of higher humidity would make those individual lifetimes irrelevant. A classic relationship in thermodynamics[10] requires that warmer air can hold more water vapor, meaning that on average, a warmer atmosphere will be a wetter one. This is a positive feedback to climate change.

Precisely how does the Earth's temperature track CO_2 concentrations? A widely used metric is the *climate sensitivity*, the equilibrium[11] temperature change associated with a doubling of atmospheric CO_2. Although estimates vary, a loose consensus is that climate sensitivity is about 3 °C. Translated into the more familiar numbers of CO_2 concentrations, an increase from 250 ppm (pre-industrial) to

---

[8] We will see in Chapter 6 that there are subtleties associated with this presumed relationship, and that in the Western Mountains, the equation "warmer = more fire" is far from universal. Globally, on a warmer and possibly wetter Earth, the future of this equation is also uncertain.

[9] But this feedback could be expected to interact with albedo, in that increased forest biomass may lower the albedo of the land surface, giving a positive feedback to climate change.

[10] More technically, the *saturation vapor pressure* in the atmosphere, a measure of its water-holding capacity, depends nonlinearly on the temperature, by the Clausius–Clapeyron equation. Consequently, for every increase of 1 °C in temperature, the water-holding capacity of the atmosphere increases by about 7%. Note to the careful reader: this appears to be a linear increase because there is another non-linearity between SVP and water-holding capacity.

[11] Here we have to remember the Earth's energy balance. The realized temperature change associated with climate sensitivity will not happen all at once, because of the different lagged responses in different parts of the Earth system to an energy imbalance. The rate and shape of how the change occurs are reflected in a *climate response function*, which measures how the fraction of the total response to a climate forcing increases over time. Typically, and intuitively, this function is *concave down*: the slope decreases with time since the forcing is initiated.

500 ppm (expected mid-twenty-first century in most projections) means an eventual global temperature increase of 3 °C.[12] Most of this increase would occur by 2100.

"Climate is always changing". Indeed, there is periodic temporal variation in Earth's climate, both in cycles that vary over 5–6 orders of magnitude[13] and with no obvious regularity over larger orders, up to the lifetime of the Earth (~4.5 × 10[9] years). Despite the regularity of seasonal cycles, the nonlinear nature of the Earth's climate system, and the "short memory" of the atmosphere,[14] virtually guarantee that no two consecutive years will be alike. Because of the thermal properties of water versus air, memory in the oceans is longer than that of the atmosphere,[15] producing climate variation on scales of several years to decades, exemplified by the El Niño Southern Oscillation (ENSO), the Pacific Decadal Oscillation[16] (PDO), and other longer (multi-decadal) quasi-periodic variation.

At scales of many thousands of years, climate responds to changes in the Earth's orbit around the sun. *Milankovich cycles* produce latitudinal, seasonal, and overall variation in solar energy received by the Earth, through changes in the tilt of the Earth's axis (a 41,000 year cycle), precession[17] of the axis (26,000 years), and eccentricity[18] of the orbit (413,000 years). Because of interactions among the orbital parameters, and feedbacks within the Earth system, climate does not cycle regularly

---

[12] If this increase occurs, Earth's temperatures would be their warmest in the last 35 million years.

[13] The expression "order of magnitude" refers to numerical differences in powers of 10, or equivalently, adding or subtracting zeros to whole numbers. So for example, one million (1,000,000 or 10[6]) is six orders of magnitude larger than 1. For example, with respect to climate, temperatures can vary in both 100-year and 10,000,000-year cycles.

[14] Technically, the memory of a process refers to how far back in its history events are influential. A sequence of rolls of dice has zero memory, because previous results have no effect on future outcomes. The thermal inertia in the ocean means that the ocean has a longer memory than the atmosphere.

[15] More precisely, it is the interactions between the dynamics of the oceans and atmosphere that operate on the lower "frequencies" of the ocean's thermal dynamics. For details see K.E. Trenberth and J.W. Hurrell. 1994. Decadal atmosphere-ocean variations in the Pacific. *Climate Dynamics* 9:303–319.

[16] It has been claimed that the PDO is neither decadal nor an oscillation. The terms are approximate and heuristic. By no means does it change like clockwork, or swing back and forth with the regularity of a pendulum. But there do indeed appear to be two opposing states, and they switch at roughly decadal intervals.

[17] Precession is the change in orientation of the axis of a rotating body. The classic example is watching the top of a spinning gyroscope make its own circle (much more slowly than the gyroscope spins).

[18] The Earth and other planets have elliptical orbits, as deduced by Johannes Kepler in the seventieth century. The eccentricity of an ellipse is a number between 0 and 1 that describes its departure from circularity (0 is circular). The precise calculation involves energy and mass considerations beyond pure geometry, but for our purposes it is enough to note that a change in eccentricity changes the *perihelion*, or closest point to the sun in the orbit, and the *aphelion*, or farthest point. The complications do create "subcycles" of 95,000 and 125,000 years, adding "noise" to the dominant 413,000-year cycle.

at any of the frequencies of the Milankovich cycles.[19] For example, a particular combination of tilt (its maximum), precession (Earth's axis inclined toward the sun at the summer solstice), and eccentricity (Earth at its multi-millennial perihelion) maximizes the solar energy received by the Northern Hemisphere in the summer, maximizing heating of most of the Earth's land mass and melting of snow and ice.

Regional variation in weather is apparent to us all, but the regional effects of climate change are harder to detect than the global ones. For example, California has a multi-year drought while the Northeast suffers a spate of unseasonably cold wet winters. A reasonable person may see the connection easily between California's weather and global warming, but the Northeast seems to be going the opposite way. We know that the climate forcings, from greenhouse gases and aerosols, are acting on a complex system (Earth's), generating interactions and feedbacks at different scales. These scales are constrained partly by landforms,[20] which are essentially fixed, and by atmospheric and oceanic circulation,[21] which are highly dynamic but nonetheless fall into observable spatial patterns. Changes in the strength, position, depth, and direction of circulations will change regional climate, without necessarily changing global averages. The increased complexity associated with regional-scale feedbacks and interactions makes for greater uncertainty in detecting and projecting regional climate change than for the global domain.

How Do We Know It?

Given what seems to be variation on every scale imaginable, in patterns varying from apparently regular oscillations to apparently pure noise, how did we detect the signal of global warming, and how do we know what it will do in the future, for which we have no observations?

A common misconception, even among some scientists, is that our understanding that the Earth is warming comes from running incredibly complex simulation

---

[19] And of course, the greenhouse effect of changing atmospheric CO_2 concentrations, whether "natural" or anthropogenic, must be overlain on these orbital forcings to estimate the Earth's energy balance at any time in its history. For an exquisite graphical presentation of the Milankovitch cycles, see the Wikipedia page, https://en.wikipedia.org/wiki/Milankovitch_cycles, and scroll down to the graphics starting at "Axial tilt".

[20] Landforms are the masses and shapes of everything from continents, through mountain ranges, to local topography.

[21] For example, *Hadley cells* are tropical atmospheric circulations that generate (at different spatial scales) hurricanes, trade winds, and jet streams. The *thermohaline circulation* is an oceanic circulation driven by gradients of temperature ("thermo") and salinity ("haline"). Both of these are highly dynamic but stable, at least currently, in their global positions.

models that all disagree and have large uncertainties.[22] The models do inform our understanding, as we shall see below, but the basic physics of the Earth's energy balance and the greenhouse effect have been understood for over 100 years. Informed by that basic physics, researchers have looked closely at empirical data sets—both climate observations from instrumental records and reconstructions of climate from the past—for evidence[23] that the physics is being realized as expected in a complex environment.

The best indicator of global warming is something that is difficult for individuals to perceive: an upward trend in global average temperature. CO_2 is *well mixed* in the atmosphere, meaning that it disperses rapidly around the Earth. Because concentrations are essentially the same everywhere,[24] the greenhouse effect is measured best at the global scale. The difference between global average temperatures in 1850 and 2015 was 1 °C (1.8 °F), reflecting greenhouse warming.[25] For those of us who barely notice the difference between 68 °F and 70 °F in the same day, let alone as an annual average, this may seem like a tiny amount. To put it in perspective, one has to look at the past, for both the departures from long-term averages and how quickly things changed, then versus now.

Paleoclimatology is the use of paleological methods to study the Earth's past climate. Although past values of many variables have been reconstructed, at many spatial and temporal resolutions, what concern us here are reconstructions of temperature and CO_2 concentration. These variables, and their correlations over time, provide context for the current rate and magnitude of global warming.

The best known temperature reconstructions are from tree-ring analysis, or *dendrochronology*. Trees at their upper elevational limits often have temperature as

---

[22] For example, the renowned physicist Freeman Dyson states "My first heresy says that all the fuss about global warming is grossly exaggerated. Here I am opposing the holy brotherhood of climate model experts and the crowd of deluded citizens who believe the numbers predicted by the computer models. Of course, they say, I have no degree in meteorology and I am therefore not qualified to speak. But I have studied the climate models and I know what they can do." (https://en.wiki-quote.org/wiki/Freeman_Dyson). Dyson goes on to describe things that the models do well and not so well, but misses the point that the models themselves are not the basis for concluding that the Earth is warming.

[23] Many have observed that a good theory, whether basic or complex, is good no longer when confronted with contradictory data. And it is more tractable to prove a hypothesis wrong than to prove it right. As William James said, "In order to disprove the assertion that all crows are black, one white crow is sufficient."

[24] But this does NOT mean that temperatures are increasing at the same rates everywhere. The many climate forcings and feedbacks do vary spatially, such that warming takes a different pace both latitudinally and regionally. For example, the polar regions are warming about twice as fast as the lower latitudes.

[25] For these and other related data, see the 2015 installment of NOAA's "State of the Climate" report, at https://www.ametsoc.org/ams/index.cfm/publications/bulletin-of-the-american-meteo-rological-society-bams-state-of-the-climate/. Also in 2015, the atmospheric CO_2 concentration passed 400 ppm (briefly, in summer) for the first time in more than 100,000 years. As I write, though, it is now expected to stay above 400 ppm for decades, at least, even if emissions are curtailed sharply.

their principal limiting factor. The classic dendrochronological reconstruction, of which there are many world-wide, identifies old trees expected to be limited by temperature, then extracts as long a record as possible of annual ring widths from increment cores or (more rarely) wedges cut from the trees.[26] Meteorological records, dating as far back as they are reliable, are obtained for as close to the tree-ring sites as possible. One then has annual time series of growth (ring widths) and temperature, of which the growth series is much longer: centuries to even millennia as opposed to decades. One or more *statistical models*[27] identifies an association between the two series during the (shorter) period of the temperature record, and then that association is extrapolated into the past to the limit of the tree-ring series, giving estimates of annual average temperature (or whatever other variable is being reconstructed) for a much longer time than that covered by the meteorological records.[28]

The well known "hockey stick" graph demonstrates how dendrochronological reconstructions indicate the unprecedented nature of recent global warming, going back 1000–2000 years.[29] The first hockey-stick graph was based entirely on tree rings, but later reconstructions were based on more diverse *proxy records* (data that substitute for a variable of interest, in this case temperature). In all these cases, besides the evidence that the late twentieth century was the warmest period for millennia, the "blade" of the hockey stick shows that the sustained rate of temperature increase through the twentieth century is unprecedented.[30]

The Earth's history is six orders of magnitude longer than the hockey-stick reconstructions. We have to turn to other proxy records to reconstruct temperatures further back in time. The further back we go, the poorer the temporal resolution becomes, and the greater the uncertainty of all reconstructed values. For our pur-

---

[26] We are skipping many of the details of dendrochronological methods here to concentrate on their objectives. A good resource for more information is the University of Arizona Laboratory of Tree-ring Research: http://ltrr.arizona.edu/.

[27] Correlation, discussed in Chapter 1, is the simplest statistical model. Other commonly used statistical models are *analysis of variance* and *linear regression*. Detailed discussion of the difference between simulation models and statistical models would take us far afield, but they are loosely contrasted by the (jargony) terms *forward modeling* and *inverse modeling*, respectively. See Chapter 4.

[28] The alert reader may see this as a classic case of the proverbial "tail wagging the dog". If climate (temperature) causes or limits growth, shouldn't we be using climate to predict growth, and not the other way around? It turns out that statistics like these work in both directions if one is cautious about inferring relationships between cause and effect. In the case of reconstructions, we are *calibrating* temperatures, making numerical estimates without attributing causes.

[29] See https://en.wikipedia.org/wiki/Hockey_stick_graph for the graph and more discussion.

[30] And our alert reader here may suspect that we are comparing apples and oranges. The twentieth century record is instrumental, and earlier values are reconstructed. A statistical model based on the former is used to estimate the latter, and there is always uncertainty around the latter. The late twentieth century observations are above the uncertainty ranges of all the statistical estimates, however, reinforcing their unprecedented nature. For those who haven't seen the hockey stick graphic, go to my favorite resource. https://en.wikipedia.org/wiki/Hockey_stick_graph (but be careful with a naive search or you'll get sports gear).

poses, two major lessons come from the monumental and diverse efforts of paleo-climatologists to write the history of the Earth's climate.[31]

First, there were periods during which annual temperatures were far above what they are today. For example, the post-dinosaur Earth, known as the Cenozoic Era, is divided into 7 *epochs*. We have referred already to the last of these, the Holocene. At the transition between the first two was the Paleocene-Eocene Thermal Maximum, or "PETM", around 55.5 million years ago. Paleological data show two pulses of carbon release into the atmosphere,[32] increasing global average temperatures by 5–8 °C, at least five times the change from the pre-industrial period to now. During this period there was a large turnover in the microfossil record, and new orders or mammals, including primates, appeared in Europe and North America.

Second, the rapidity of current change is unprecedented. The temperature "excursion" in the PETM was brief at geological time scales, "merely" about 200,000 years, but the average annual rate of carbon emissions was an order of magnitude slower than it is today.[33] Given the fairly rapid turnover of the biota in the PETM, the proportional consequences for the current era are alarming.

The most visible tools in climate-change science, to the general public and to scientists in other disciplines, are the general circulation models, or "global climate models" (GCMs).[34] GCMs are *simulation models*: they replicate the physical processes that are thought to be the most important to represent accurately in order for the model *output* (in this case spatial and temporal patterns of temperature, precipitation, wind, and many other variables) to be realistic, believable, and robust to the many uncertainties involved in modeling. Every few years, the Intergovernmental Panel on Climate Change (IPCC) produces an update, or "Assessment Report"

---

[31] Besides dendrochronology, the tools of paleoclimatology for inferring temperatures include (1) ice cores taken from the ice caps of Greenland and Antarctica, with analyses of trapped air, changes in the thickness of layers within the cores, and the ratio of heavy (atomic weight 18) to normal oxygen isotopes; (2) analysis of sediments, which may preserve fossils, biomarker molecules, chemical signatures such as the magnesium/calcium ratio in *foraminifera* (primitive microorganisms related to protozoa), and oxygen- or carbon-isotope ratios. The oldest ice cores reach 1.5 million years, or two orders of magnitude longer than tree rings. Useful sediment records for paleoclimatology date back at least 200 million years.

[32] These were detected by carbon *isotope analysis*. The ratio of $^{13}C/^{12}C$ was "suddenly" much lower in marine and terrestrial carbonates and organic carbon. See P.L. Koch, *et al.*, 1992. Correlation between isotope records in marine and continental carbon reservoirs near the Paleocene/Eocene boundary. *Nature* 358:319–322.

[33] The rate during the PETM carbon pulses is estimated to be between 0.3 and 1.7 Pg (10^{15} g) of carbon per year (PgC/year). The figure for 2015 is 9.795 PgC/year. For the PETM estimate, see Y. Cui, *et al.*, 2011. Slow release of fossil carbon during the Paleocene-Eocene Thermal Maximum. *Nature Geoscience* 4:481–485. For the 2015 estimate, see https://www.co2.earth/global-co2-emissions. PgC/yr and "gigatonnes" per year are equivalent.

[34] These two terms that produce the same acronym are somewhat interchangeable, although some will argue strongly for one or the other being right or wrong. I will use the acronym throughout, for which it is good to keep in mind both definitions. The first refers to the underlying physics of directional motion in the atmosphere; the second refers to the spatial scale (global), with its associated strengths and weaknesses for simulating the Earth's climate.

(there are five so far and a sixth underway), on climate change. A core activity associated with these reports is to run a group of these models, or *ensemble*, and to compare, contrast, and combine the outputs.[35] The results of these efforts are ranges of predictions and levels of confidence about them.

The "simplest" GCMs (and even these have thousands of parameters) simulated just the atmosphere. As computing power has increased, and modelers from different disciplines have convened and shared their work, processes in the models, and in some case the models in their entirety, have been *coupled*. Across models, processes affect each other, and there are feedbacks. For example, the Earth's oceans and atmosphere interact continually. The effects of ENSO (anomalies in sea-surface temperatures) on climate are some of the best known, but interactions occur on temporal scales from nearly instantaneous to centuries-long.[36] These interactions are represented in "coupled atmosphere-ocean GCMs", or AOGCMs. Extending the idea, "Earth-system models" aim to simulate all the important interactions in the Earth system, including the biological and even the social in some cases. As I write, however, these most complex models are eschewed in favor of the physical models (GCMs and AOGCMs) for climate-change modeling.

GCMs take an existing state and let it evolve according to the laws of physics. This is both a strength and limitation. It automates an immense sequence of calculations whose parameters have been thoroughly tested, like following a complicated road map from point A to point B. On the other hand, if the existing state has been mis-specified, or if the physics is imprecisely represented, the models are not self-correcting. Consequently, much of the time and effort spent to evaluate GCMs has involved simulating historical periods for which we have good observations and measuring the discrepancies between the models and "reality" at the end of the observational record. For example, a GCM that simulates twentieth-century climate with and without greenhouse forcing should match observations in the first case but not in the second. Without that match, we should have no confidence that the model can project the climate's response to greenhouse forcing in the future, for which we have no observations to compare with model output.

Let me emphasize that the principal value of climate models is not to identify the causes of global warming, nor to detect its signal. We have the aforementioned basic physics and paleological and historical records that do that. Climate models (and all simulation models, by definition) let us produce "if, then?" scenarios for the past,

---

[35] The term *ensemble* also refers to repeated runs of the same simulation model. Because many components of these models are *stochastic*—they are represented probabilistically and therefore differently for each realization—multiple runs of a single GCM, or one or more runs of several GCMs, produce a range of outputs whose statistics may make a "safer" statement about results than a single model output. Consider throwing a pair of dice once, or many times and taking the average. Which gives a better estimate of the expected result?

[36] Almost instantaneous, as in the time it takes for colder water to change the temperature of warmer air. Over centuries, heat from greenhouse warming will build in the deep ocean, and its transfer to average global temperatures of the atmosphere will lag far behind rising CO_2 levels.

present, or future. Given a state of the Earth system, and our understanding of the physics, what climate could we expect?

How Well Do We Know It?

Information is a difference that makes a difference.
   —Gregory Bateson

The most noticeable aspects of climate are the extreme and the sudden, but the best understood are the global and gradual. For example, we understand the connection to climate change much better for global average temperature than for an unprecedented tropical cyclone. Our immediate perceptions detect single anomalies better than changes in long-term averages.[37] Conversely, though, the statistics produced by a wealth of observations are more reliable than a single event for predicting the future. Even these are far from ideal, however.

Consider the case of average global temperature. As I said earlier, our observations of its gradual increase are the strongest evidence of "global warming" (more or less by definition). In the ideal, to support our inferences, we would have a blanket coverage of the Earth, at regular spacing, of instruments that would record temperatures at regular intervals for decades. What we have is far from this ideal. Because most of the Earth's surface is ocean, and much of the land surface is remote and costly and time-consuming to reach, instrumentation is spaced very unevenly. It turns out that a simple average cannot be estimated directly (for various technical reasons), and must involve a statistical model. The essence of these models is to use a *weighted average* of existing observations, in which observations that are more reliable, or represent more surface area, than others are given more influence in the calculation.

Global surface temperatures are a reflection of the Earth's energy imbalance that has been realized. The unrealized part, which will appear over centuries, is the heat content of the deep ocean. Until recently, we had few measurements of the deep ocean, so climate models make many assumptions about *ocean mixing*, how quickly heat trapped by greenhouse forcing is dispersed into the ocean depths. For example, if this rate is over-estimated, then simulated surface temperatures will be under-estimated, a discrepancy that will increase as projections are made further into the future.

The observations with the greatest uncertainty are aerosols, the second most important greenhouse forcing after CO_2. Because CO_2 is well mixed, high-quality measurements at individual locations are a good proxy for global levels. In contrast, aerosols often come from point sources and their concentrations are linked to atmospheric circulation, but not in ways that can be estimated easily. There is great

---

[37] For obvious reasons, in that our species evolved to notice things like the sudden appearance of a predator more clearly than a change in annual temperature of a few degrees.

potential for integrated use of satellite observations to produce the analogue of the modeled global temperature record, but as I write[38] the estimates of aerosol forcing in climate models are still partly *ad hoc*.

The limitations to paleoclimatology are of two types. The first is *record length* (how far back in time do we have observations?). Clearly we have a longer record here than for instrumental observations, so one could argue that anything helps. In general, what we can infer becomes less *precise* (more variation or uncertainty around estimates, such as average global temperature) as we go back in time, and inferences have poorer resolution (e.g., an observation represents a century, or a millennium, rather than a year or a decade). The second is that correlations between temperature and the metrics we so carefully measure are sometimes weak. For example, correlations[39] between ring widths of trees and temperatures are rarely above 0.5, "half" of a perfect correlation. Despite these limitations, multiple lines of evidence in the paleoclimatic record point to the climate forcing from CO_2 that we use to infer the speed and magnitude of global warming and inform our GCMs.

These GCMs are among the most complex of human creations. Typically, they have literally thousands of *adjustable parameters*,[40] to which model results can be very sensitive. These parameters are components of the nonlinear equations that simulate global circulation patterns and are the core of GCMs. Two principal limitations are associated with these parameters; both contribute to the uncertainty of outcomes. The first is their sheer number, and the cumulative effect of many of them being "not exactly right". The second is the sensitivity of some of the simulated processes[41] to very slight changes even in a few parameters. This is the so-called "butterfly effect", a property of nonlinear dynamical systems such as the atmosphere.

These limitations are intrinsic to models as complex as GCMs, and are unlikely to disappear even with expected advances in atmospheric science, oceanography, or computer science. Researchers compensate for the limitations in two ways. First, models are compared to observations in many different ways. For example, if a GCM is "correct", should it not be able to replicate the climate of the twentieth

---

[38] My most recent source is the Hansen *et al.* (2011) paper referenced in Note 6.

[39] Recall from Chapter 1 ("terms") that a correlation is the simplest statistical comparison, and assumes a linear relationship between two variables. In deciphering the paleoclimatic record, other more involved statistics are common. For example, often two variables change in what appear to be rough cycles, with one lagged behind the other, as temperature often lags behind changing CO_2 concentrations. This pattern will produce weak linear correlations, but still allows for one variable to serve as a proxy for the other.

[40] Recall our definition of a parameter in Chapter 1, as something that should have a particular value, but we may not know what it is. Those that we know are called fixed. For example, in Einstein's famous equation, $E = mc^2$, c is the speed of light, which we know to several decimal places. Those parameters that we don't know are called *adjustable*: they are "tuned" based on observations, or theory, or both.

[41] Here it would not be unreasonable to ask "but aren't we interested in the real processes, not the simulated ones?" I would argue (and there are many philosophical subtleties that go way beyond the scope of this book) that science is intrinsically a process of building models of reality, though clearly in a much different way from how fiction does it. Real clouds, or real oceans, don't have parameters at all, only their models do.

century, for which we have reliable observations? This simple prescription is complicated to implement, because it involves "tuning", or *calibrating*, the right combinations of the adjustable parameters in a model such that the output matches observations satisfactorily.[42] A big assumption is required here. It is that one did not get the right answer for the wrong reason; for example, as in "two wrongs make a right", two biases introduced by poor calibration canceled each other out. If so, the model is fragile to new conditions, such as the future. Testing models against paleoclimatic observations can offset the problem partly, but not completely.

A second route to compensating for the limitations to GCMs invokes the proverbial "making lemonade from lemons". Researchers accept that all models have intrinsic limitations and turn the associated uncertainties to their advantage. Almost all working models are now *stochastic*,[43] in that simulations are repeated, several to many times, with parameters deliberately fixed at different values. The output from these *ensemble* simulations is used to put realistic bounds on the possibilities, rather than to predict an outcome (e.g., future) that is exactly right. In practice, ensemble modeling takes place at two scales: parameters within individual GCMs[44] are varied, and different GCMs are run with the same sets of initial conditions, and these multi-GCM ensembles put (even) wider ranges on future possibilities.

A final dependency (more than a limitation) of GCMs is they depend upon ongoing input from societies in the form of GHG and aerosol emissions. For example, atmospheric circulations depend on the heat fluxes to and from the Earth's surface, which are sensitive to radiative forcing, as we saw above. Modelers have used three different approaches for estimating these emissions, but seem to have settled on "Shared Socioeconomic Pathways" (SSPs), which specify the cumulative amount of climate forcing by 2100 from different scenarios of human-caused emissions. These range from "negative emissions" (causing the Earth to capture more GHGs than it emits to the atmosphere), to "business as usual" (no change in rates and trends of emissions). Using the range of SSPs, ensembles of GCMs produce widely different estimates of annual global temperature increases in the future.[45]

---

[42] "Satisfactory" could take many forms here. Typically, it is the *aggregate properties* of observed climate—averages, trends, ranges of values—that are compared, rather than attempting to match single observations, e.g., for a day, or for a single location. The latter constitutes *false precision*, essentially answers that are better than one has a right to expect, and which therefore are suspect.

[43] I am taking liberties with the term "stochastic", by using just one of its many meanings. Formally, a "stochastic process" is distinguished from a "deterministic process" in that some elements of the former are considered to be truly random. This raises the question again of the distinction between reality itself and models of reality (see Note 40). By some definitions, there are no processes in nature that are truly stochastic, or random, above the quantum level (and maybe not even there), whereas there are many stochastic models of processes in nature. Fortunately, we need only consider a subset of the latter here.

[44] As I write there are many GCMs, the products of modeling groups around the world. There have been two "Coupled Model Intercomparison Projects", associated loosely with the IPCC work but implemented separately. For the most recent, "CMIP5", see http://cmip-pcmdi.llnl.gov/.

[45] For example, by 2100, ensemble projections (from many GCMs) predict 1.0 °C increase with negative emissions, but 3.7 °C increase with business as usual.

Overcoming the limitations to climate models is less tractable than for those noted above to observations or paleoclimatology. In particular, a recent comparison[46] of the uncertainties among models from the last IPCC report (AR5) vs. the previous (AR4) suggests very little change in the breadth of outcomes, even though many advances in the science had been incorporated in the more recent versions of the models. We will probably have to live with the "lemonade" of projections for the foreseeable future.

How Will It Affect the Western Mountains?

If you don't like the weather, wait ten minutes.
    —attributed to local wisdom in many mountain climates

In the GCMs that serve us well in projecting future climates, entire mountain ranges can be represented by just a few homogeneous grid cells.[47] Anyone who has experienced mountain weather can testify that the action is at much finer scales. We study mountain meteorology at spatial scales from the "micro" to the *synoptic*[48] or "meso" scale. We know that over distances as short as tens of meters (less than 100 ft.), mountain *microclimates* can be different. Even without shade from trees *per se*, topographic shading can make a difference, for humans, between uncomfortably warm and uncomfortably cold. Over 100 s of meters, the differences, such as those between ridgetops and valley bottoms, can be equally striking.

The key sources of climate and weather variation at the mesoscale in western mountains are *orographic* (specifically referring to the topographic relief of mountains). Orographic variation is nested within the global latitudinal gradient of temperature from Equator to poles and large-scale variation in atmospheric circulation,[49] and it interacts directly with the altitudinal gradients in mountain ranges. Basic knowledge of mountain meteorology is well established.[50] Some of

---

[46] R. Knutti and J. Sedláček. 2013. Robustness and uncertainties in the new CMIP5 climate model projections. *Nature Climate Change* 3:369–373.

[47] GCMs are typically at a *horizontal grid spacing* (roughly the lengths of the sides of rectangular cells) of 50–100 km, sometimes larger, within which key elements, like topography, temperature, and snow depth, are represented by a single number. For perspective, the direct distance (as the crow flies) from Mt. Whitney (14,500 ft. elevation) to Badwater, Death Valley (−286 ft. elevation) is only 138 miles. In some models, they would be in the same grid cell; in most others they would be in adjacent cells.

[48] Synoptic-scale meteorology, literally from the Greek word meaning "seen together", is a large scale when considering typical meteorological forecasting, but smaller than the global scales of GCMs. The phenomena associated with it in standard forecasting include the high- and low-pressure systems seen on weather maps, weather "fronts", and storms such as extra-tropical cyclones.

[49] This larger-scale variation is what is represented best in global climate models.

[50] For example, the best book I have come across, for both completeness and readability, is R.G. Barry, Mountain Weather and Climate, Methuen Publishers, 1981. One can also find a contemporary view with the always informative Wikipedia, at https://en.wikipedia.org/wiki/Category:Mountain_meteorology.

the key elements are the interactions of temperature, moisture, and wind on a landscape of varying relief, from rolling hills to cliffs, canyon walls, and *cirques* (dead ends of canyons) that can be thousands of feet high. These interactions are constrained partly by the laws of physics, but subject to uncertain variation in a changing climate. For example, *cold-air drainages* (convective movement downhill of cold air in watersheds), leading to temperature inversions, are an inevitable outcome of orographic dynamics, and well understood physically, but their strength is contingent on synoptic weather.

Ideally we would estimate the changes in microclimate and meso-scale climate in mountains with no more uncertainty than our estimates for global-scale climate. For example, will the average temperature of a mountain valley in 2100 be more or less warmer than today than the global average will be? Unfortunately, the uncertainty in our future projections increases as we go to finer scales. As complex as the GCMs are, they can still make use of spatial averages, such as the average temperature and elevation of a 100-km grid cell, and simulate large-scale processes, such as atmospheric circulation or heat exchange between atmosphere and ocean, reasonably accurately. The downside of such large-scale predictions is we have estimates only at those scales.[51] We can assume that these averages hold for smaller areas only at our peril, but as we downscale from GCMs to regional (mountain) climate and then to microclimate, we end up relying on averages more and more.

Climatologists have two distinct methods for downscaling GCMs for regional and finer-scale projections. Briefly,[52] they are *dynamical* and *statistical* downscaling. Dynamical downscaling simulates atmospheric processes analogously to GCMs, but at much finer grid spacings (as I write they are down to 1 km). Some processes that are too detailed to be included in GCMs, such as aspects of cloud dynamics and convection, are at the "right" scale for *regional climate models* (RCMs). RCMs rely on large-scale averages in that they are constrained by *boundary conditions*: typically values from GCM simulations that ensure that RCM results are not biased (i.e., their averages very different from the GCMs'). These constraints can be fairly simple or very complex.

Statistical downscaling, in contrast, does not really simulate anything. Instead, it takes advantage of observed fine-scale variation and superimposes it on GCM-scale averages. The simplest way to do this is the *delta method*. For what you want to predict, e.g., daily (or annual) temperature, monthly rainfall, etc., take the difference between the GCM-scale average that corresponds to your observations and the one that corresponds to the time period you want to predict. If the future value is larger,

---

[51] For example, we may have perfectly good estimates of the average hourly temperature across the Northern Rocky Mountains, or the Sierra Nevada, and its effect on the calculation of global average temperature, but this is not enough if we want to understand the future climate of a highly valued mountain watershed that is thought to be vulnerable to global warming.

[52] The details can become very technical very quickly. For a much fuller discussion, see an open-access article: D. McKenzie, U. Shankar, *et al.*, 2014. Smoke consequences of new fire regimes driven by climate change. *Earth's Future* 2:35–59. It is about fire and smoke, but gives an in-depth overview of downscaling GCMs.

add that to the observations to get the answer.[53] If it is smaller, subtract it from the observations. One key difference from dynamical downscaling is that in theory, statistical downscaling can be applied "all the way down" to microclimate, whereas the limits to dynamical downscaling are coupled to the limits in modeling atmospheric processes. In practice, however, climate-change projections with both methods are generally considered useful only to the resolution of RCMs. We will see in later chapters how uncertainties compound also in ecological studies as they interpolate or extrapolate across scales.

We are now ready to ask what we know about how climate may change in the western mountains. We have two sources of specific information. The first is empirical research that asks if mountains have characteristics, in general, that cause us to expect a different response to global climate change than we see in global averages. The second is regional climate modeling that projects changes for specific mountain ranges. Together they can give us good information, though not exact predictions.

In the first category is *elevation-dependent warming* (EDW), literally increases in temperature at higher elevations that are greater than the global average.[54] There is fairly good evidence that in some regions, and in some seasons, EDW has occurred already and will continue to do so.[55] If we are to be confident that EDW is a relatively stable phenomenon[56] that is robust to details of Earth-system dynamics, at least in the current era, we should be able to identify its various causes.

The EDW working group identified several physical mechanisms the make sense as causes for EDW. Albedo and aerosols are both climate forcings on a global scale, as we saw at the beginning of this chapter. They are also the clearest drivers of EDW. With the loss of snow cover at middle and upper elevations that has happened just with warming temperatures to date, the surface albedo in mountains has decreased, thereby decreasing reflectance and increasing the proportional absorption of incoming radiation. Similarly, most of the aerosols that dim incoming radiation are concentrated at elevations below about 9800 ft. (3000 m), such that the high elevations of the Rocky Mountains and Sierra Nevada in particular receive more

---

[53] For example, suppose I wanted to predict the average winter temperature in the northern Rocky Mountains for 2060, at 1-km resolution, so that I could report to land managers what to expect in 40 years. In the delta method, I would take the average difference for the (very few) cells in whatever GCM I was using, and add (or subtract, if it's negative) it uniformly to the values for the (many) 1-km cells of interest. For those interested in statistics, the averages for the two time periods would be different, but their variances, or standard deviations, would be the same.

[54] Recall that depending on the GCM and the specific socio-economic scenario(s) used, these global averages vary, along with GCM-scale projections for different regions.

[55] An international collaboration of 21 researchers convened in 2015 to provide an overview of EDW and some of its probable causes. I have drawn heavily on their paper for this discussion. Mountain Research Initiative EDW Working Group, 2015. Elevation-dependent warming in mountain regions of the world. *Nature Climate Change* 5:424–430.

[56] Recall Chapter 1 in which I express a hope that this book will remain current as long as possible in its explanations, if (obviously) not in the details of ongoing research post-publication. To do that, I focus here and elsewhere on physical and biological relationships that should hold as the climate continues to warm.

(unblocked) radiation than do lower elevations. These local effects are separate from the global greenhouse effect of CO_2, and also from the latitudinal gradient of warming from the tropics (less) to the poles (more).

We are now ready to ask what particular climate changes are expected for the Western Mountains, based on both EDW and regional-scale climate projections from simulation models. The summary "predictions" below inform Chapters 4–7, in which I focus on particular ecological phenomena; Chapter 8, in which I look at some of the associated complexities of these phenomena; and Chapter 9, in which I look at the human dimensions of climate change. In the process, we need to be mindful of the uncertainties associated with the climate projections themselves, and with our discussion in Chapter 1 of the constants vs. the dependencies in global change.[57]

Based on what we know about the mountains (Chapter 2) and the climate projections (this chapter), we can expect two general themes to emerge, one for the northern "half" of the western mountains, and one for the southern "half". More details and some of the consequences follow in Chapter 4. Chapters 5–7 explore the ecological consequences. Very briefly stated, in the North water changes from snow to rain, but in the South it goes away (Table 3.1).

In the North (Cascade Range, Pacific Coast Ranges N, Northern Rocky Mountains), freezing levels rise, more precipitation falls as rain and less as snow, and what snow there is melts earlier in the year. Total available water (for ecosystems and humans) may not decrease, although with earlier snowmelt and runoff its "timing" will be inconvenient.

Table 3.1 Changes in mountains relative to global climate

Mountains	Temperature increase	Precipitation	Comments
Cascade Range and Pacific Coast Ranges (N)	Less	More winter, less summer	Maritime influence
Rocky Mountains (N)	Same		Glacier loss
Sierra Nevada	Same or more	Less	Snow loss
Coast Ranges (S)	Same or more	Less	Drought
Rocky Mountains (S and central)	Same or more	Less	Snow loss and drought
Sky Islands and Basin Ranges	More	Less	Drought

Relative temperature change in this table does not incorporate EDW

[57] These summaries draw almost entirely on the 2018 4th National Climate Assessment, Volume II. https://nca2018.globalchange.gov/, which evaluates the effects of climate change in the United States, at present and through the twenty-first century. Assessments appear every four years, on average.

The Cascade Range and the Pacific Coast Ranges (to the Klamath and Trinity Mountains)

Except for the volcanoes, these mountains and their glaciers are at lower elevations than many of those in the Interior West. Mass loss from glaciers will be substantial, more rain will fall and less snow, making "rain-on-snow" events more frequent, with greater risk of flooding in the surrounding lowlands. Seasonal snow will melt earlier in the year, and the elevation of the perennial snow line will rise. The good news (relatively) for the Pacific Northwest is that the projections are for less warming than in other areas of the West; the maritime climate will serve as a buffer, at least for a few decades, against warming.

The Northern Rocky Mountains

Changes will be similar to those in the maritime climates to the west, but in this more continental climate, the buffering from the maritime influence is not expected. Glaciers will shrink or disappear except for those at the highest elevations.

In the South (Sierra Nevada, Pacific Coast Ranges S, Southern and Central Rocky Mountains, Sky Islands and Basin Ranges), warming is expected to be exacerbated by increased drought, particularly in the inland Southwest, including persistent droughts lasting for several years to a decade or more. In these already arid climates, total water availability could be much less, as well as the timing of snowmelt and runoff being earlier.

The Sierra Nevada

Multi-year droughts, at least as bad as the 2012–2017 drought, are expected. Perennial snowpacks, which even at the highest elevations did not always persist in dry years of the twentieth century, may become rare instead of the norm. The 13 glaciers here are all at above 11,500 ft., and are expected to persist for at least a few decades.

The Pacific Coast Ranges (South of the Klamath and Trinity Mountains)

Snow is not a major force in these low-elevation mountains even now, except for the few areas above ~7000 ft. in the Transverse Ranges. As the climate warms, it will become even more marginal, and perhaps be reduced to an occasional dusting.

The Southern and Central Rocky Mountains

These are some of the highest mountains in the West, with many summits above 14,000 ft. Snow may persist at high elevations, though for less of the year. At lower elevations, it will shrink or disappear, with associated effects (more seasonal, see Chapter 4) on river- and stream-flow.

The Sky Islands and the Basin Ranges

These mountains will likely be affected by drought the most severely and in the most different ways. The isolated Sky Islands and the more extended Basin Ranges will see their characteristic mountain environments (forests and alpine) shrink or disappear.

In the chapters that follow, we will see how our "constant" themes from Chapter 1—snow and ice, evolution, movement, disturbance, and interaction—have consequences for key elements of western-mountain landscapes: water, vegetation, animals, and people. In each discussion we want to be aware of the particular "stage", or stages (which mountains), on which the consequences are acted out.

Chapter 4
Water Towers of the West

Water is the driver of Nature.
—Leonardo da Vinci

The Western Mountains are one of the most active components of the global water cycle. Between snow-covered mountain tops and the flatlands below, gravity is the driver of water's passage over the land surface. The four great rivers of the West, the Columbia, Missouri, Colorado, and Rio Grande, accumulate the output of snowmelt and rainfall to the thousands of smaller rivers, streams, and creeks that make up the hydrological networks of the West. So the changes in the amounts and timing of the volume of water in these rivers, a resource vital to society, depend on broad-scale geographic patterns of snowmelt and rainfall. Moving upstream through their tributaries, the amounts and timing depend more and more on local or short-term weather. Storms bring flooded streams, at lower elevations from sheer volumes of precipitation and at higher elevations exacerbated by *rain on snow* events (either untimely warm periods in winter or heavy spring rains on snowpack).

Moving "back" one segment in the global water cycle,[1] the water that leaves the West by land arrives by air (Fig. 4.1). So the amounts and timing of arrivals depend on interactions between the oceans and the atmosphere. Evaporation from the oceans is the main source of water for that atmospheric circulation.[2] The global average rate of evaporation is an element of the Earth's *hydroclimate* (literally, water's part in climate) that may be sensitive to global warming, because evaporation requires heat, of which there will be more, on average. In turn, this may

---

[1] For a good overview of the global water cycle, and more graphics, see the USGS web page at https://water.usgs.gov/edu/watercycle.html. I shall be referring to the cycle in subsequent chapters in many different ways.

[2] The oceans hold almost all of the Earth's water: about 96.5%.

© Springer Nature Switzerland AG 2020
D. McKenzie, *Mountains in the Greenhouse*,
https://doi.org/10.1007/978-3-030-42432-9_4

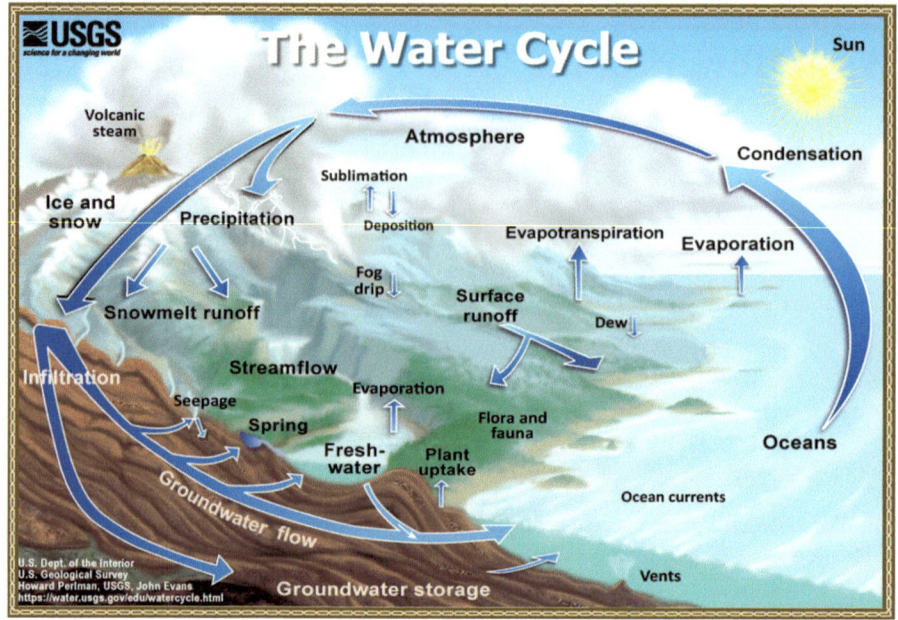

Fig. 4.1 The complete water cycle. Graphic courtesy of the USGS

"intensify" the global water cycle by increasing the rates of other elements, such as precipitation and runoff.[3]

The climate system, which we discussed in Chapter 3, includes the movement of water through the atmosphere, driven by *thermodynamics* (the forces driven by energy) and *fluid dynamics* (the forces associated with constrained movement, such as in liquids). To understand potential changes in the global water cycle caused by a warming climate, we have to understand how both types of forces respond to increasing inputs of heat to the system. The increased evaporation adds water to the atmosphere, so that more is available for precipitation, and that water may be re-routed by the climate system into different patterns of circulation.[4] For example, the region where most of the Earth's rain falls is called the *Intertropical Convergence Zone*, or ITCZ, on average about 6° (latitude) north of the Equator. The ITCZ seems to adjust position in response to differential warming of the two (N&S) hemispheres, moving toward the warming hemisphere, which itself has dispersed heat toward the

---

[3] It would take us too far afield—mostly away from the mountains—to go into the rather complicated logic of estimating "intensity". I will refer to some particulars later in the chapter, with reference to "capacitance". For the interested reader, a good overview of the effects of warming on the water cycle is in T.G. Huntington, 2006. Evidence for intensification of the global water cycle. *Journal of Hydrology* 319:83–95.

[4] How this actually happens is quite complicated, and is one of the biggest sources of variation in the many GCMs on whose improvements thousands of scientists spend their lives. We will be content here to note some of the broad-scale consequences for the West.

cooler hemisphere, maintaining an energy balance. This movement of heat affects the *Hadley Circulation* (one of the broad-scale fluid dynamic processes in the atmosphere), thereby shifting the ITCZ.[5] This illustrates how fluid dynamics and thermodynamics interact in the climate system.

For tropical ecosystems then, movement of the ITCZ can make a big difference. Its seasonal movement creates the dry and wet seasons typically associated with the tropics. How it might shift in response to climate change is presently unclear. In theory, the Northern Hemisphere may warm faster than the Southern, at least in the near future, because of loss of Arctic ice cover, but as I write this is work in progress for the GCMs. But in any case, the ITCZ's wildest excursions have not taken it as far north as the mountains of the West.

Other changes in circulation, both regional- and fine-scale, will have more immediate consequences for the Western Mountains. The Southwest has suffered through multi-year or multi-decadal droughts in the past, and much of the region had extreme drought in the summer of 2018, the 18th year of a persistent drought. This long history plus GCM simulations suggest that "climate-change drought" is a distinct possibility[6] for the Southwest in a warming future. Regional-scale projections for the rest of the West are more variable, so the region that is currently the hottest and driest will be hardest hit by a changing hydroclimate.

Many of us[7] think of "the monsoon" as something that begins over the Indian Ocean and strikes the Asian continent in mid-summer, but North America has its own, albeit weaker than the Asian version. From July to mid-September, afternoon thunderstorms occur often across the Southwest, although the peak of monsoon rains is farther south, in the mountains of northern Mexico. The monsoon is a regional process, but the spatial patterns of rainfall are local. The *convective activity* (literally how water is "pulled" through the atmosphere) associated with thunderstorms varies with topography, so that monsoon rains are hard to forecast in the mountains of the Southwest even in a stable climate. We expect, however, that the North American monsoon may be a victim of global warming. Using a *regional climate model* (a finer-scale cousin of the GCMs), researchers recently projected[8] much less rainfall in the northern part of the monsoon (i.e., over the Southwest), adding an extra insult to projections of increased drought overall for the Southwest.

---

[5] For a detailed technical explanation of this, see https://climate-dynamics.org/why-does-the-itcz-shift-and-how/. Briefly, the ITCZ is an area of low pressure where two Trade Winds meet just north of the Equator.

[6] This phrase was coined by David Breshears and colleagues in a 2005 journal paper: "Regional vegetation die-off in response to global-change-type drought". *Proceedings of the National Academy of Sciences, USA* 102 (42) 15144–15148; https://doi.org/10.1073/pnas.0505734102. Put simply, the hotter it is, the worse it is (for vegetation and ecosystems) to be dry.

[7] OK, particularly those of us who have followed Himalayan climbing over the years and know that the Indian monsoon signals the end of the spring window for any serious mountaineering.

[8] S. Pascale, *et al.* 2017. Weakening of the North American monsoon with global warming. *Nature Climate Change* 7:806–812.

At a finer scale, the intensifying global water cycle increases the likelihood and frequency of *atmospheric rivers*, literally flows of concentrated moisture in the atmosphere. The so-called Pineapple Express, forming near the Hawaiian tropics and striking the west coast of the U.S., is the best known example, but atmospheric rivers are continuously formed around the globe; as many as five in each hemisphere at any time. They are typically several thousand kilometers long and a few hundred kilometers wide, many carrying more water than the Amazon River. Atmospheric rivers can produce extreme rains and flooding, because they carry about half of the total annual rainfall and create nearly half of the annual snowpack in the mountains in their path.[9] GCMs will never be able to forecast individual atmospheric rivers,[10] but simulations, guided by the intensification of the water cycle, suggest that there will be more of them. Besides atmospheric rivers, *convective storms* (i.e., with thunder and lightning) are the other finer-scale precipitation extreme. These are also difficult to represent in GCMs, but some recent work (2017)[11] suggests that the total rainfall from convective storms in the USA will almost double by the end of the century. As is the case today, atmospheric rivers will be more a feature of maritime climates, and convective storms of continental climates.

The Magic Line: Snow or Rain, Frozen or Melted

The Earth's climate has seen far more severe extremes than anything during human history. From the Hadean period, ending 4000 Mya (4 billion), with an average surface temperature of 230 °C,[12] to the so-called "Snowball Earth", sometime before 850 Mya, when Earth's surface was nearly entirely frozen,[13] to the 150 million-year sojourn of the dinosaurs, average annual temperatures have differed by hundreds of degrees Celsius. More recently, during the Paleocene-Eocene Thermal Maximum

---

[9] Remembering Chapter 2, this means the Cascade Range and the Sierra Nevada, as well as all the smaller ranges mentioned in "The Pacific Coast Ranges."

[10] In Chapter 8, I will talk about extreme events and the limits to our forecasting power for them, but we will see that forecasting their general patterns and frequencies is much more feasible than predicting individual events.

[11] A.F. Prein, *et al.* 2017. Increased rainfall volume from future convective storms in the US. *Nature Climate Change* 7:880–884.

[12] Yes, that's 446 °F, way above the boiling point of water at *normal atmospheric pressure*. But there was so much CO_2 back then that atmospheric pressure was 27 *atmospheres*, i.e., 27 times what it is today.

[13] Some have argued for the less definitive term "Slushball Earth", because if we accept that there was really a hard freeze, and that it may have happened more than once, it's difficult to explain how the planet could have warmed back up enough to melt its frozen surface. In this latter case, water remains around the Equator, maintaining the global hydrological cycle and the associated thermodynamics that permit re-warming. See W.B. Harland, 2007. Origin and assessment of Snowball Earth hypotheses. *Geological Magazine* 144:633–642.

(or PETM), ~55 Mya, annual average temperatures were likely about 8 °C hotter than today. During the Ice Ages, beginning about 2.6 Mya and which we are technically still in,[14] they varied from about 5 °C lower than today[15] at the peaks of glacial periods to roughly the same as today during the "interglacials".

In contrast, during the current Holocene Period,[16] although there have been warm and cold spells, global temperatures have been unusually (for Earth) stable. Moreover, it could be said that we are in a "sweet spot" of climate for optimal benefits for human civilization: not too hot; not too cold. What does that mean?

The key *parameter* for the sweet spot in climate is the freezing point of water.[17] Whether precipitation falls as snow or rain, and when or whether it melts if it is snow, makes a big difference for everything that happens in mountain ecosystems. One of the key issues in both weather (forecasting) and climate (projections) is figuring out what will fall.[18] In the broadest sense, a climatic sweet spot for civilization is where it is cold enough, on average, for there to be enough snow to keep rivers and streams flowing all year in temperate latitudes, but warm enough that it melts to produce that water, rather than having "too much" of it frozen year-round. Because the Earth's temperature varies so much between the tropics and the polar regions, the extremes of the temperature *gradient* from Arctic to Equator need to be within the range in which human survival (and that of organisms, plants and animals, on which we depend) is possible. The subset of that range represented by Holocene climate is thought to be optimal.[19]

With our focus on the Western Mountains, we want to know where and when snow will change to rain. More specifically, for any location, with its unique elevation, latitude, and topographic position, what times of the year that now or formerly had snowfall will have rain in the future? In Chapter 3 we noted the limits to the

---

[14] Here some have argued that we would be heading into a glacial period right now were it not for the "first" anthropogenic global warming caused by the advent of agriculture. That discussion would take us too far afield, but the key scholarly work in that area is W.F. Ruddiman, 2003. The anthropogenic greenhouse era began thousands of years ago. *Climatic Change* 61:261–293.

[15] And I must qualify "today" to mean what temperatures would be without human-caused climate change.

[16] The Holocene is technically the interglacial period that we are in. I will leave discussion of the "Anthropocene", claimed by different factions to have started anywhere from the beginnings of agriculture to practically yesterday, to others.

[17] Sneak preview: A key *variable* that we will encounter below is the *water-balance deficit*, a plant-relevant measure of how dry things are.

[18] And of course how much, but that is a different problem. For those interested, the jargon term is "phase partitioning", referring to *algorithms* (computational procedures) that use combinations of average, minimum, and maximum daily air temperature to produce equations for either the snow fraction (during precipitation) or snow frequency.

[19] To my knowledge, no one has put together a comprehensive study of why this is the optimal range. The point is often made very broadly, with reference to the Ice Ages and the Age of the Dinosaurs as being outside such an optimal range. It is fairly clear why an ice age would be a struggle for civilization. At the other end, an Earth that is too hot on average will make some regions more livable, but make more area unlivable and reduce the productivity of agriculture, particularly in arid regions.

precision of climate projections for the future, and how the limits are more constraining at finer scales in space and time. For example, just as we cannot predict the high temperature on April 15, 2040, in Leadville, Colorado, we cannot predict the date of the last snowfall of the year (although April 15 would not be a bad guess). For snow and rain, we can compromise between information that is too general to be useful or is too precise to be believable.[20]

The *snow water equivalent* (SWE) of any mass of snow is the depth of the water (over the same area) that would be left if all the snow melted. So it is a combination of average snow depth and snow density. It turns out that SWE has a strong correlation with temperature, stronger than with precipitation. Observational data on SWE in the Western Mountains over the past few decades provide a check on the projections of SWE by climate models and *ecohydrological models* (see the end of this chapter). As long as we are not seeking unreasonable precision, and are content with such estimates as "average SWE on April 1 in the 2040s", SWE is a good variable for identifying the "line" between snow and rain.

Another useful tool is the *hydrograph*, literally a graph showing the rate of flow, or *discharge*, over time, of moving water at some location. As the West warms, the total volume will reflect the throughput of precipitation, the *intensification* of the global water cycle that I referred to above. More important for our purposes, however, seasonal patterns will change, notably *peak discharge* happening earlier in the year, reflecting earlier snowmelt in a warming world (See Note 20).

Hydrographs that aggregate streamflow from many sources are the most strongly associated with temperature changes, so the hydrographs of the four great rivers of the West mentioned above are good indicators of when and how much snow melts (or is rain to begin with) over wide areas. By comparing their hydrographs, we can see the cumulative effects of such factors as the elevations of their headwaters, regional-scale temperature change, and changes in the spatial patterns of global circulation affecting both seasonal and total precipitation. For example, we saw in Chapter 2 that the Cascade Range is mostly at lower elevations than the other great mountain ranges of the West, but I said in Chapter 3 that warming in the maritime Pacific Northwest is projected to be less than the average, both global and for the West. These factors will offset each other, in that there will be less "movement" overall of the line between snow and rain, but more area may be vulnerable to the consequences of a crossing of the line. Future hydrographs of the Columbia River (both simulated and observed) could suggest the dominant factor, by comparing them to hydrographs[21] of the Colorado River, for example, whose headwaters are at higher elevations, but in a region expected to warm more than the Pacific Northwest.

---

[20] For examples of hydrographs, see https://en.wikipedia.org/wiki/Hydrograph.

[21] For this question, we would compare the seasonal or monthly proportions of the total flow rather than the totals themselves. The latter would be better evidence for changes in the global water cycle. For example, if peak flows moved more (to earlier) in the Columbia than the Colorado, we might infer that even though warming in the Northwest was below average, a lot of acreage was crossing a threshold of melting, because of the relatively low elevations in the Cascade Range.

SWE and hydrographs give us the before and after of the snow vs. rain dynamic, but SWE can also be treated like a hydrograph. In snow-dominated landscapes across the West, the greatest SWE is often around the beginning of April. When SWE starts to decrease—and it does not always do so *monotonically* (i.e., without reversing itself)—it means simply that more snow melts than falls. With warming, this time comes earlier in the season, on average, just as peak discharge on a hydrograph moves earlier.[22]

As I noted in Chapter 1, the freezing (melting) point of water (ice) does not vary with climate change. That means that in a *non-stationary* climate (remember our definition) our "magic line" has to change position, both in latitude and altitude. In most cases this will be to higher elevations and toward the poles. One of the clearest places to see these changes is in the glaciers of the West, to which we turn next.

Glaciers: A Pure Signal of a Warming Climate?

Eight states in the West have named glaciers.[23] Washington leads the most recent count with 186, followed by Montana (60), Wyoming (37), Oregon (35), California (20), and Colorado (16). Nevada and Utah have one each, but Nevada's is usually considered a *rock glacier* (see Note 16, Chapter 2), and Utah's a "buried" glacier.[24] As you might expect, our glaciers track the *Laramide Belt* and the *Nevada Belt* (recall Chapter 2), becoming smaller and fewer from north to south.[25]

What defines a glacier is that it is a mass of ice that is heavy enough to move (under the force of the Earth's gravity). Most glaciers move slowly—they are often said to "flow like paint"—with a velocity that is controlled predominantly, though not completely, by the slope of the terrain on which they lie.[26] In a perfect equilibrium, ice would accumulate at the upper end of a glacier and melt at the lower end. The *mass balance* of a glacier is positive when it is growing in mass (and therefore

---

[22] It may occur to the astute reader that one could make much out of comparing hydrographs with SWE measurements from the right locations upstream. True, in theory, but among the logistical difficulties of such a comparison is establishing a rate of "snow-cover depletion" that can be compared statistically with the hydrograph. Matching the spatial pattern of observed or computed snowmelt with the necessarily fixed pattern of stream hydrographs is another issue.

[23] I am indebted to two web resources for the numbers given in this section: "Glaciers of the American West" (http://glaciers.research.pdx.edu/index.html) and the ever-useful Wikipedia.

[24] Just for completeness, the rock glacier in Nevada is on Wheeler Peak, Nevada's highest point and in Great Basin National Park. Utah's *buried glacier* (literally, under the land surface) is Timpanogos Glacier, in the Uintah Mountains, Utah's highest, in the northeast corner of the state. Neither of these is much of an indicator of climate change.

[25] But not so much as you might think. That is because elevations of the ranges increase from north to south.

[26] For example, the Khumbu Icefall on Mt. Everest moves a meter (over 3 ft.) per day, on average, but most move at least an *order of magnitude* more slowly.

volume in lockstep with mass[27]) and negative when it is shrinking (thus zero at that perfect equilibrium. Of course more ice accumulates in the winter, and more melts in the summer, thus every glacier has an *accumulation zone*, and an *ablation zone*, in which these processes are occurring, respectively. A key element of glaciers for us is the *equilibrium line altitude* (ELA), the elevation mid-glacier that separates the two zones. During the year, the ELA moves predictably (up in summer, down in winter). Over longer periods, the average ELA will also move predictably in response to climate, to the extent that the dominant controls on glacial mass balance are climatic. In that case, a warming climate causes more ice to melt in a year than to accumulate, on average, than would occur in a stable climate.[28]

If a glacier's ELA moves permanently above the upper boundary, a glacier is doomed. This is something that is not easy to detect directly, because it involves some complicated computations, but it is inferable from the visible shrinkage of glaciers around the West (Figs. 4.2 and 4.3). Many glaciers are retreating upward in elevation alarmingly quickly. For example, at current rates, the glaciers in Glacier National Park will disappear by 2040, with the exception of the Piegan Glacier, which will last longer partly due to its being the highest (8200–8800 ft. asl.)[29] (Fig. 4.4). Other less iconic glaciers around the West are similarly threatened, particularly those close to our magic line, by virtue of lower elevation or above-average projected warming.

Loss of glaciers is one of our purest indicators of a rapidly warming climate. Just like the climate itself, however, not every instance supports the overall trend. Analogously to the unusually fierce winters in the northeastern USA in recent years, not every glacier is retreating in our globally warming climate. *Surging glaciers* (literally, glaciers that sometimes flow much faster than their average rate) can be deceptively out-of-sync with annual climate, and can cause "excursions" from their expected pattern of advance or retreat. This brings complications to attempts to project the fate of all glaciers, but also falsifies claims that an advancing glacier, or a few, is evidence against a globally warming climate.

Glaciers may be changing rapidly in our current climate, with its rate of warming that exceeds that of any other warming phase in the last 65 myr,[30] but their change is still slow enough that it is the slowest part of the global water cycle. In the context of global climate change, melting ice is a slow (but positive) feedback, thereby

---

[27] The density of glacial ice is stable for all practical purposes, hence the lockstep. By contrast, snow-water-equivalent (SWE) varies greatly, so that snow's volume is a poor surrogate for its mass.

[28] Even in a stable climate, glaciers can change size from year to year and decade to decade, for many reasons that are outside our scope here. As discussed in Chapter 1, the *detection* of changes and their *attribution* are separate tasks, though complementary.

[29] I am grateful to my colleague Dan Fagre at the USGS in Glacier National Park for summarizing a lot of computation and projection for me in this information. Not every glacier is measured every year, so our "current rates" are based on periodic repeated photography (see Figs. 4.2, 4.3, and 4.4).

[30] For example, recall from Chapter 3 (Note 32) that our carbon release is 10× that of the PETM, the largest excursion (upward) from the mean climate of the Cenozoic Era.

Fig. 4.2 Grinnell Glacier (**a**) 1941, photo by J.A. Donnan, (**b**) 2013, photo by K. Jacks. All glacier photos are courtesy of the USGS and Glacier National Park

a

W.C. Alden Photo, GNP Archives, 1913

b

Agassiz Glacier, Aug. 2, 2005
Greg Pederson photo, USGS

Fig. 4.3 Agassiz Glacier (**a**) 1913, photo by W.C. Alden, (**b**) 2005, photo by Greg Pederson

a

b

Fig. 4.4 Piegan Glacier (**a**) 1938, photo by T.J. Hileman, (**b**) 1998, photo by Lisa McKeon

accelerating warming, but not immediately. Similarly, glaciers in the Western Mountains will slow mountain-ecosystem change in response to warming, relative to what will occur in the non-glaciated ranges. Glaciers serve as *capacitors*,[31] "storing" the effects of the Earth's energy imbalance,[32] though temporarily. This storage is a buffer against rapid warming, and as such it favors the more northern ranges over their southern counterparts, which have little or no buffering. Snow cover serves a similar purpose, although its capacitance is weaker, lasting only months in mountains with seasonal snow, but like glaciers, though still more volatile, in mountains with perennial snow. With the projected differences in larger-scale hydroclimate between the South (drier) than the North (wetter or the same), and the zero capacitance in the South, the water-related effects of warming climate will be more severe in the American Southwest, the southern Sierra Nevada, and the southern Rocky Mountains.

Changes in Mountain Hydrology

Mountain hydrology includes hydroclimate, a mountain *cryosphere* (literally the "sphere" in which water is in its solid form), a network of waterways shaped by mountain topography, water storage in plants and the soil, and evapotranspiration (ET). This last term refers to the combined process of evaporation from the plant surface and the soil and *transpiration* (the respiration,[33] or breathing, of the plant, in which CO_2 is fixed through the plant's surface, during which water is lost). When we estimate the effects of a warming climate on mountain hydrology, we need to consider the direct effects of more energy input to each of these elements, and the interactions and feedbacks among them. Just as with the climate system itself, this gets complex quickly, so we will start with the simple principles, in the order of the elements above.

At the start of this chapter I discussed the effects of energy input to the hydroclimate, including the complexities of deciding whether rain or snow falls to begin with. The effects of energy input to the cryosphere are relatively simple: heat melts snow and shrinks glaciers. Similarly, increasing energy input to the waterways raises water temperatures. For our purposes here, mountain topography is impervious to energy inputs.[34]

---

[31] This is a loose metaphor, which I use because it will help us understand implications of the rates of ecosystem change in later chapters. More precisely, capacitance measures how electric charge changes with a change in voltage. Imagine (loosely, please) gravitational potential as voltage and energy imbalance as electric charge. A glacier (capacitor) "holds" the energy from falling (literally downhill) when it would otherwise flow away.

[32] Recall our discussion of the Earth's energy balance in Chapter 3.

[33] A process I will not discuss is *soil respiration*, the expulsion of CO_2 from soil organisms.

[34] In Chapter 6, I will touch briefly on disturbances involving land movement, so not completely impervious.

We now need to bring in vegetation as an element of mountain hydrology, though reserving the main discussion of vegetation for Chapter 5. There is a vast literature on plant-water interactions,[35] and vegetation will be a player through the rest of the book. More energy input to vegetation increases metabolic rates,[36] and thus the need for water, the water lost (to the plant) through ET and thus gained by the local atmosphere, and the *negative feedback* to the vegetation's need for water through evaporative cooling from transpired water. There is also a more complex feedback operating here: the plant-water interactions depend on the density of plant biomass, which depends on the *limiting factors* for plant growth. Earlier (Chapter 1: concepts) I contrasted *energy-limited* and *water-limited* systems. In energy-limited systems, plants grow faster with increased energy input. In water-limited systems, they grow more slowly because their growth is compromised by sub-optimal water availability. Many mountain ecosystems of the West are not purely of either type, or change seasonally or in dry versus wet years.

A widely used metric for the degree of water versus energy limitation is the water-balance deficit (DEF),[37] which estimates the difference[38] (usually in millimeters per unit time) between observed or calculated ET ("actual" ET, or AET) and the more rapid ET that would ensue if a unit of vegetation had all the water it needed ("potential" ET, or PET). DEF is somewhat complicated to calculate, and requires some approximations, but is useful for diverse applications. Basically, the greater DEF is,[39] the drier it is from a plant's perspective, and it turns out that one can make reasonably good predictions about ecological outcomes from it.

Lastly, recalling our discussion of hydrographs, increased energy input to the entire system changes the seasonal distribution of peak discharges <u>within</u> the network of waterways, because of earlier snowmelt. What it does to the proportion of *runoff* (water that moves downhill under gravity) that orginates <u>outside</u> the waterways is more complicated because the water that is transpired is lost from that runoff.[40]

So when we seek to put all of mountain hydrology together to project the consequences of a warming climate, we are in a similar situation to researchers studying the (non-stationary) climate system. Many of the guiding principles are well understood: the greenhouse effect for the climate system, or the melting and freezing of water in mountain hydrology, but the complexity of potential outcomes, given the

---

[35] For a technical review relevant to our issues, see N. McDowell, *et al.* 2008. Mechanisms of plant survival and mortality during drought: why do some plants survive while others succumb to drought? *New Phytologist* 178:719–739.

[36] For now, I am ignoring the effects of CO_2 fertilization on all of the plant-water interactions. Chapter 5 will address that briefly.

[37] "DEF", not "WBD", in accordance with common use.

[38] Or a normalized difference. I will use just the simple "PET-AET".

[39] DEF should always be a positive number. If it is negative, one of the approximations for either PET or AET has caused an anomaly.

[40] Not all of it, of course, because some ET recondenses and falls back to the ground.

number of variables and the interactions and feedbacks, compels us to turn to simulation models, with all their strengths and limitations, for answers.

Eco-hydrological Models: How Can We Use Them for Climate-Change Projections?

It's tough to make predictions, especially about the future.
    —Yogi Berra

In Chapter 3, I made a few observations about global climate models, the chief tool of the IPCC Integrated Assessment Reports on global climate change and its consequences. They take an existing state of the atmosphere and let it evolve according to the laws of physics. Predictions from these models for the future cannot be compared with observations, so we assume that those laws of physics were specified precisely and correctly. We take some comfort if the model replicates present-day conditions reasonably accurately, but because of the complexities of the models, such as the number of parameters that have to be specified, this is not always reliable.[41]

Eco-hydrological models (shorthand for "ecosystem simulation models that incorporate hydrology") work similarly in that they take an existing state of the ecosystem, in its broadest sense including the local atmosphere and the terrain, and let it evolve according to the laws of biology and hydrology.[42] As with climate models, the output can be as voluminous as one wishes,[43] but everyone who devotes the time and energy to running a model will want to know some standard things. For example, a "bare bones" *query* (request for output) from a GCM might be "For each cell, for each year, what are the average temperature and total precipitation?" If we want to know how much more one region warms over a decade than another, on average, and whether it is wetter or drier, and if we trust our model completely, this is enough. For an eco-hydrological model, an equivalent question might be "How much more (less) biomass is on the landscape than when we started, and how much earlier (later) is the peak discharge?" Rarely, if ever, would one go to the trouble of

---

[41] That is, there can be reason to assume that the models that replicate today's climate best are not robust to the changes, usually in interactions and feedbacks, that the future will bring. Part of this is the problem of *non-stationarity* that I introduced in Chapter 1. I know of no one-stop shopping ("further reading") that can provide a good overview of these issues.

[42] And of the fundamental sciences that underlie them: physics and chemistry. Even though the atmosphere is involved, technically, in these models, typically it is only as a source of specified inputs to equations governing the biology and hydrology.

[43] That is, one can specify in the computer program that runs a simulation model that it create a database that includes as many of the model's calculations as desired. A tradeoff occurs in that the more output you request, the longer it takes to run the model and the more data you have to sort through at the end.

running a complex model and not take advantage of at least some other general information in the output.

The scale[44] of interest is an overarching control on the design of eco-hydrological models. One can ask a model to do only a finite number of things. In eco-hydrology, this limit and the physical constraints imposed by landforms have focused simulation models into two "sizes", which I will call the regional scale and the watershed scale.[45] Regional-scale models are matched to the scale of regional climate models, and watershed-scale models to the extent of discrete stream and river networks. So I will make a simple contrast here, which we will refine through the rest of this discussion. We use regional-scale models to simulate the future of the West in a warming climate, whereas we use watershed-scale models to simulate the future of the mountains of the West. My focus in this book is on the second of these, but we will see below how regional-scale models contribute indirectly to that effort. First, however, I want to elaborate on the sorts of questions we can and should ask with models. This theme will recur in Chapters 5–7, with an extended discussion in Chapter 8.

Simulation models are an example of *forward modeling*, consisting of inferring the consequences of one state or process for another. Generally, cause and effect are assumed; the program[46] moves from the former to the latter in one or more time steps. For example, rainfall is simulated and runoff at the river mouth, or anywhere downstream, is one effect. Soil water storage is another. A hot dry summer is simulated and tree mortality is the effect. Each of these processes creates a feedback to the system. Soil moisture not lost to runoff is a buffer to water stress on plants, and tree mortality reduces the live biomass that needs the water. These feedbacks correspond to the "forwardness" of the model, although one often sees feedbacks represented by loops in system diagrams, including those for simulation models. So if we trust that we have the causality right in our forward-modeling scheme, we can reach into the future, for which we have no observations to verify that we represented our system accurately.

In some scientific disciplines, forward modeling is robust because parameters are known and *fixed* (the same in every situation). These cases can be simple and obvious, like the "2" in the formula for the area of a circle, or fantastically complicated, but still always the same.[47] In the Earth sciences, the overwhelming majority of

---

[44] Recall our discussion of scale in Chapter 1. Following that, here and in Chapters 5–7 "scale" will refer to a combination of "grain" and "extent"; i.e., "fine-scale" means finer resolution (grain) and a smaller area (extent) simulated.

[45] I am certainly not the first or only person to do this, and other divisions could be justified, and often are. My division is heuristic, for this book.

[46] This is usually a computer program, but the process itself can be more abstract. For example, one could mentally, or on the back of the proverbial envelope, or in a structured social setting, fill in the boxes and arrows of a conceptual "forward" model, and be justified in claiming that this was a simulation.

[47] *Fine-tuning* problems, a subject not for the faint of heart, are cases in which for one or more parameters there seems to be no reason why they are what they are. But just like that "2", they are always the same when observed.

parameters are *adjustable* (they are contingent on factors that vary). *Inverse modeling* is the technique for finding those contingent parameters in the situations that they are relevant. Their properties enable us to be mathematically precise in the reverse of forward modeling: inferring causes[48] from effects (consequences). For example, if we watch a predator catch prey, we infer from forward modeling that the prey will be dead, but if we find a dead animal, we have to use inverse modeling to figure out what killed it.

If you fear that inverse modeling is a lot more difficult than forward modeling, and generally inspires less confidence, you are right. So modelers should stick with forward models as much as possible? But here is the problem. In most, if not all, of our (forward) simulation models, the presumed cause-effect relationships within them are compiled from one or more inverse-modeling efforts.[49] This is not a bad thing *per se*, but it constrains how we can use models for future projections when we have no (future) data to which to compare them.

One of the most important steps in doing science is deciding what questions to ask (some would call it the most important). The value of their answers is certainly a big part of choosing questions, but for our purposes I will limit the discussion to "What questions can we answer, satisfactorily, for a warming climate in the Western Mountains?". We have our tool set, eco-hydrological models in this chapter and others for Chapters 5–7. We also know that our model includes both forward and inverse modeling, with the inverse models being less in plain view, and the forward models sometimes being a partial or even total composite of inverse models.

Let's begin with a broad guideline for choosing questions,[50] building on our discussion in Chapter 1 of what persists and what changes. This one will work for our purposes: understanding how the Western Mountains will change in a warming climate, but is also quite general. Use what is constant to project what will change. Simplistic? How will this get us to mountain ecosystems in a greenhouse-like world?

For mountain hydrology, the constant melting point of ice, or snow, is the root of our inference, as we saw above in this chapter. From this constant, the first steps in our inference are to the *phase partitioning* of precipitation (to snow or rain, see Note 18), rates and timing of snowmelt, water available for vegetation, and variation in streamflow and peak discharges. Already things are getting complex (which is why we need a simulation model). Each of these calculations is a forward-modeling *algorithm* (per Note 18) in our simulation system, with *parameters* that may have

---

[48] Recall here, of course, the phrase "correlation is not causation", which I introduced earlier. The possible discrepancies between "causes" inferred from inverse modeling and real causes can be a significant issue.

[49] I won't cite examples here, because at this stage they could just as easily confuse as enlighten. In Chapter 8 we will see how to recognize the limits, but also the benefits, of this process when we discuss the complexities of models more fully.

[50] The philosophy of science is replete with variations on the theme of how to structure a research program. For those interested, I would start with searches on keywords like "scientific method", "research paradigm", and the like. For ecological research, a good overview, though somewhat technical, is E.D. Ford. 2000. *Scientific Method for Ecological Research*. Cambridge University Press.

been chosen through *inverse modeling*, as described above. So here is the key. Which of these parameters can we trust to be valid in a warming climate?[51] Here are some pitfalls, which I will ask you to think about intuitively for now. In Chapter 8 we will revisit them in more detail, after we see some examples in Chapters 5–7.

Sensitivity to Initial Conditions, or States If we don't know an initial state precisely enough, how can we "let it evolve according to the laws of physics" as above in this section, or any other paradigm.

Things Change at Different Scales Often what we care about is at a different (usually broader) scale than what we have a good model for.

Stationarity We covered this earlier. Sometimes controls on processes (such as *limiting factors*) change.

Back at last to eco-hydrological models, those at the watershed scale are one of the more robust tools[52] that we have for climate-change research. For example, one of the most exhaustively tested is called RHESSys[53] (Regional Hydro-Ecologic Simulation System). It builds on decades of both empirical research and earlier-generation models, and limits its future projections to hydrology and carbon dynamics. It is part of a family called "big leaf" models, because it represents vegetation simply as biomass rather than trees, shrubs, and herbaceous plants. The advantage of this for climate-change projections is that states evolve according to laws that are very well known. For example, the *carbon balance* (how much the total changes over time) is based on photosynthesis, *respiration* (literally breathing, except that plants "inhale" CO_2), and *decomposition* (the breakdown of plant tissue by microorganisms). Photosynthesis in particular is "hardwired" biology; its parameters are not *adjustable*.

Climate is part of the *boundary conditions* of the model (initial states imposed from outside the model proper). Part of these can be produced by a regional-scale eco-hydrological model, or a regional climate model, or both, one of whose cells may include the entire domain of the watershed-scale model. This is a one-way process with watershed-scale eco-hydrological models, with no feedbacks from the finer-scale model to the boundary conditions.[54] RHESSys, for example, has 19

---

[51] I am avoiding a more subtle and complex issue for now: what about the structure of said inverse models, as opposed to just the values of individual parameters? These models often are assembled by complicated statistical procedures that can have many pitfalls along the way and large *confidence intervals* (ranges of plausible values other than the accepted ones).

[52] In this author's opinion.

[53] A lot of these models seem to acquire tongue-twisting names. RHESSys is a descendant of the model BIOME-BGC. A regional-scale relative of RHESSys is VIC (Variable Infiltration Capacity [Model]). For a definitive description of RHESSys, see C.L. Tague and L.E. Band. 2004. RHESSys: Regional Hydro-Ecologic Simulation System—An object-oriented approach to spatially distributed modeling of carbon, water, and nutrient cycling. *Earth Interactions* 8:1–42.

[54] Some models used for climate-change projections do include feedbacks from the land surface to the atmosphere.

"environmental inputs" that serve as boundary conditions, including climatic variables such as minimum and maximum temperature, precipitation (rain + snow), relative humidity, and wind speed.

It is now time to state explicitly an overarching constraint on all the models that are relevant to our topic. Models don't answer the question "what will be…?", they answer "what if…?" Believing that we are asking and answering the former leads to all sorts of trouble.[55] With ensemble climate models, which I discussed in Chapter 3, we are asking "IF socioeconomic scenario X (or RCP Y, or Shared Socioeconomic Pathway Z) were to be realized, what would be the range of outcomes for Earth's future climate?" With an associated eco-hydrological model, which is contingent[56] on those outcomes, we ask "IF that climate were to be realized (providing boundary conditions for our model), what would be the range of outcomes for the ecology and hydrology of our watershed?"

What Do We Expect for the Western Mountains?[57]

What can we say about our mountain ranges specifically? Drawing partly on Table 3.1 and the text below it, and some experiments with eco-hydrological models, below are some ideas. I will limit these thoughts here to the water cycle, not repeating the statements about the individual ranges from Chapter 3. In Chapter 5 we will see the strong *coupling* (science jargon for things influencing each other) between vegetation and water, producing much uncertainty and spatial variation, not just among regions but among and within watersheds.

In general, we expect the changes in the water cycle to be clearer and more pronounced in snow-dominated systems (the north) and in maritime vs. continental climates, because in maritime climates with their mountains being lower, the snow-vs-rain demarcation will place proportionally more territory below it (i.e., snow cover could be restricted to the highest elevations). If temperature increases are largest in late winter and spring, as expected, this will amplify the effects, and the differences, because those are the seasons with the most "action" in the *cryosphere*. At the same time, the maritime regions, especially the Northwest, may be more

---

[55] And not just scientific trouble. This is a classic error of perception in the world at large. For example, a former member of Congress and frequent guest on the news media is wont to say (about climate models) "All the predictions are wrong!" But climate-change projections are not arithmetic problems.

[56] Even if we accept the contingency, it may seem fair to object that well, the derivative model is only as good as the climate projections. "Garbage in, garbage out." But that misses the logic. The derivative model still answers our questions. What will be compromised with "garbage in" is the value of the answers for any further inference, or for policy or management.

[57] As I write, the best technical summary I know of this is C.L. Tague and A.L. Dugger. 2010. Ecohydrology and climate change in the mountains of the western USA—a review of research and opportunities. *Geography Compass* 4:1648–1663. The general observations below draw on this article.

complacent (here meaning less reactive, or fragile, rather than self-satisfied) to the effects of warming climate on the water cycle, because being wetter overall can be a buffer against the effects of summer drought.

Broadly speaking, our Magic Line will shift northward, with every part of the West expected to see some warming, annually and seasonally. Some mountain ecosystems will lose their *capacitors* (see above and Note 31), when glaciers mostly or totally melt and seasonal and perennial snow mostly or completely disappears. Where this is the principal buffer against water *shortage* (here referring to basic ecological function, not human use[58]), effects will be more severe. For example, one could reasonably argue that in the Pacific Northwest, loss of glaciers and perennial snow except at the highest elevations would not make streams and rivers run dry, whereas it very well might in the southern Rocky Mountains. As I said earlier, the mountain hydrology of the West that will be the hardest hit by expected climate change will be in the Southwest (into which I include the southernmost Sierra Nevada). In the southern regions, the Water Towers will start running dry.

---

[58] This is not to dismiss human use as unimportant. The expected changes will have major implications for infrastructure and personal water consumption in the Northwest, as elsewhere. I will touch on this in Chapter 9, but leave detailed analyses for other authors.

Chapter 5
Trees, Forests, and Carbon

The woods are lovely, dark, and deep
But I have promises to keep
And miles to go before I sleep
And miles to go before I sleep.
—Robert Frost

Forests cover somewhat less than half of the West.[1] California has the most forested land (about 20.8 million acres), but Idaho (about 20.4 million acres) has the highest percentage of forest (about 38%), followed by Oregon and Colorado. In the mountains, however, the percentages of forest are much higher.[2] Some mountains are forested from top to bottom, whereas others may have both upper and lower *treelines* (literally, elevation bands, often fairly wide, that mark a transition between a landscape with trees and one without). For example, in the Cascade Range, most mountains other than the volcanoes can have trees growing both at or near their summits[3] and at their bases. In contrast, in the central and southern Rocky Mountains and other arid ranges, upper treeline can be as high as 11,000 ft., and lower treeline as high as 7000 ft.

In the Western Mountains, almost 80% of the forests are *coniferous* (cone-bearing trees, sometimes called "needle-leaf") rather than *broadleaf* (i.e., real leaves instead of needles). This is unusual worldwide for temperate or tropical forests,

---

[1] So how much exactly? There is no one correct answer, because there is no clear dividing line between forest and non-forest. The terms "closed forest", "open forest", "woodland", and "savanna" are used world-wide to refer broadly to varying levels of tree cover, but these have their own gradations, and even the coarse-scale thresholds are not agreed upon universally. In keeping with this uncertainty, I will use the terms but not define specific thresholds between them.

[2] Once again, no precise numbers exist. A further complication is that not everyone will agree as to whether such things as "rolling hills", mesas, bluffs, etc. should be included in mountain acreage.

[3] Depending on how rocky and steep those summits are, of course. See Chapter 2.

© Springer Nature Switzerland AG 2020
D. McKenzie, *Mountains in the Greenhouse*,
https://doi.org/10.1007/978-3-030-42432-9_5

which are predominantly broadleaf.[4] The dominance of conifers may be due to several factors, including, notably, (1) a combination of seasonal patterns of precipitation and relatively mild winters that favors conifers,[5] and (2) the relatively recent (in geological time scales) colonization of the West by vegetation, after the last glaciation.

Forest Biology and Ecology: What Persists and What Changes

"The overarching control on tree distributions is climate." I have heard and read this in many places over the years, and I expect that some readers have too. We have several ways of knowing this. If we zoom out from individual trees and what makes them grow or fail, live or die, we see regional, continental, and global spatial patterns in the associations, often *correlations*, between climate and the tree species, *genera*, and *life forms*[6] at any location. We can also see temporal patterns. The paleological ("paleo") record lets us overlay reconstructed climate (see Chapter 3) and two useful indicators of tree presence and growth.

Fossil pollen and *macrofossils* (literally, fossil parts of organisms that are large enough to identify them),[7] primarily from lake sediments, indicate what trees were present[8] at or near the collection site, and give us some idea of relative abundance. The dating of these records has decadal accuracy at best, but because that matches

---

[4] The Earth's *boreal* (referring here to high latitudes, not boreal vs. *austral* for the Northern vs. Southern Hemispheres) forests, in Alaska, Canada, Scandinavia, and Siberia, are primarily coniferous. There seem to be physiological limits to "how cold you can go" for broadleaf trees. A scholarly review of this is in C. Körner *et al.* 2016. Where, why and how? Explaining the low-temperature range limits of temperate tree species. *Journal of Ecology* 104:1076–1088.

[5] More technically, if the most favorable seasons for biomass accumulation via photosynthesis are when only conifers have foliage, then they have a strong competitive advantage. Mild winters, particularly in the Cascade Range in which conifers dominate almost totally, and hot dry summers across most of the West, during which growth partly or wholly shuts down due to water scarcity, both favor a life form that can add biomass year-round.

[6] Scientific (in pseudo-Latin) names comprise the *genus* (singular of genera) and species, along with a "subspecies" if it exists. The main taxonomic hierarchy, for those who don't recall high-school biology, is Kingdom, Phylum, Class, Order, Family, Genus, Species (whose mnemonic is King Philip Cried Oh For Goodness Sake, or its less printable varieties). Unfortunately for simplicity, the most important broad distinction between tree *life forms* (a term I will use loosely throughout within trees and among trees, shrubs, grasses, and herbs) is *gymnosperm* vs. *angiosperm*, most simply cone-bearing vs. flowering. This is a division "between" Kingdom (plants) and Phylum, into six *phyla* of gymnosperms, but eight "groups" (and 84 classes) of angiosperms. For our purposes, "conifer" versus "broadleaf" covers the essentials.

[7] Sometimes we are lucky enough to identify them down to the level of species, but often it is a coarser level of the taxonomic hierarchy. The whole field of molecularly based fossil identification (from fossil DNA and other components) is outside our scope of discussion here.

[8] The aphorism "Absence of evidence is not evidence of absence" is relevant here. We can say, more or less for sure, which species or life forms were present; we cannot say which were not.

the accuracy of the best paleoclimatic records, approximately, we can use them together to infer relationships between trees and climate.

A second paleo method, *dendrochronology*, or tree-ring analysis, looks at correlations between annual growth in tree diameters as recorded in the widths of rings and climatic *variables*, such as temperature and precipitation.[9] Tree rings give us much greater precision than lake-sediment records (annual vs. multi-decadal), at the cost of having a shorter record overall (hundreds and in some cases several thousand years versus tens of thousands of years). Both are useful for understanding the climatic controls on trees, and therefore forests. Contemporary research enhances the correlative value of the paleo research, such that we can use what persists to understand what changes. For example, if we do a controlled experiment in which we measure, in detail, the physiological response of a plant (tree) to increased temperature, or a shortage of water, or an abundance of CO_2, or some combination of those, we have a biological basis for inferring broader-scale influences of climate on trees.[10] In the larger (ecological) context of forests in a changing climate, these are explored most usefully as *limiting factors*.

The two broad categories of limiting factors for our purposes are energy and water, as I observed in Chapter 1. They are heuristic because we can think of all the forests of the Western Mountains as lying on a gradient between purely energy-limited and purely water-limited. In the coldest environments, above the upper treeline, there is not enough heat, literally, to sustain metabolism minimally for survival. This is the purest case (on the landscape, not in the lab) of energy limitation.[11] Below lower treelines, in contrast, the limiting factor is water. Trees can survive most of the hot temperatures encountered in the West, if they have water: the oasis effect. Everywhere else (for our purposes, forested Western Mountains), forests (trees) can be energy- or water-limited.

Some nearly pure examples of one or the other limiting factor in ecosystems are squarely within forests. Recall the descriptions of mountain vegetation in Chapter 2. At the highest elevations on the west side of the northern Cascade Range, mountain hemlock (Fig. 5.1) is the dominant tree species. We know that it is energy-limited because in tree-ring studies the worst years for growth were those that had a heavy persistent snowpack, as measured by snow-water equivalent (SWE) in April. Across the West, usually at the lowest elevations that are still forested, a dominant species

---

[9] "Of course some do go both ways", as the Scarecrow said. We saw in Chapter 3 that tree-ring analysis can be used to infer climate, and here I am saying that it can be used to infer things about trees. With the caveat that correlation is not (always) causation, we prefer to start any analysis by being explicit about which inferences we want to make, and why.

[10] The translation across scales can be problematic, as can other extrapolations outside the domain that an initial inference was made. For example, if a bit more CO_2 is a good thing, would a lot more be better? We will look at some of these limitations explicitly in Chapter 8, and I will be careful here to avoid the "if a little is good, more is always better" error.

[11] But there is virtually no "pure" anything in ecological systems. For example, a principal cause of seedling death at high elevations is desiccation, i.e., water limitation.

Fig. 5.1 Mountain hemlock on a ridgeline in the Cascade Range that sees 7–8 months per year of snow cover. Photo by the author

is ponderosa pine. We know that it is water-limited because its best years for growth are wet (the years overall or those with wet growing seasons).

This seemingly simple polarity between limiting factors becomes complex quickly in the context of ecosystems, even those so "simple" as to be dominated by a single tree species. For example, if we follow mountain hemlock and its ecology to the southern part of its range, the growth response to warm vs. cold winters reverses itself, and it acts like a water-limited species. In most mountain forests of the West, however, multiple species co-exist,[12] and we have to think about their *interactions* to see how limiting factors are operating. Although some interactions among individual trees are *symbiotic* (cooperative or mutualistic),[13] most are competitive: in a world of limited resources (all ecosystems), what is available to one

---

[12] Though far fewer than in the East. Large areas of the Western Mountains are dominated by one or a few conifer species. Broadleaf species other than quaking aspen are usually found at the lower elevations of forests (e.g., oaks) or in recently disturbed (e.g., red alder, bigleaf maple) or *riparian* (along rivers or streams) areas (e.g., cottonwoods, alder, ash).

[13] For example, in most stands of conifers, individual trees are connected belowground by *ectomycorrhiza* (literally, fungi ["myco"] on the surface ["ecto" as opposed to "endo"] of roots ["rhiza"]). These *symbiotic* fungi enable individual trees to share resources through a connected root system. There are also claims, of varying credibility, of more profound communication among organisms through ectomycorrhiza. I will leave the interested reader to explore this.

organism is therefore not available to others, and the more organisms or the larger the organisms, the stronger the competition.

We can link broad-scale climatic controls on trees to fine-scale growth, mortality, and reproduction by observing that individual trees compete for our two limiting resources ("factors"), water and energy. Energy comes in two forms: radiation, as light for photosynthesis and heat to drive metabolism, and nutrients, which embody chemical *potential* energy.[14] Water comes mainly from the soil, through roots, rather than being absorbed directly from the air (from rain). In *open-grown* stands of trees (where competition for resources is minimal), the climatic limiting factors can be seen more clearly,[15] whereas in dense closed-canopy forests),[16] their signal is filtered through competition. Limiting factors persist in their importance, however, so we can hold them constant in projecting forest changes in a *non-stationary* climate.

All species require some of each of the limiting resources. At upper and lower treelines, we often see a single dominant species. In between, some species are nearly dominant, some rare. How do they sort this out?

Species composition in ecosystems is a vast area of research. For our purposes, we need to note that a fantastically complex evolutionary game is being played, which has persisted for hundreds of millions of years[17] through global climates more extreme than ours, now and at least for the near future. On average, a more diverse species composition suggests more complex tradeoffs in adaptation. For example, a "simple" adaptation for cold hardiness can suffice for an upper-treeline species to out-compete others. In the more species-rich forests of the Western Mountains, subtle differences in many factors can give a species an advantage, or a handicap. Along with cold hardiness, other differences are in light requirements (energy), levels of soil nutrients (energy), and drought tolerance (water). In projecting responses of forests to a warming climate, drought tolerance will be important both directly in response to climate and indirectly in the context of competition. For example, in a water-limited system, water may be more available to a tree that has fewer competitors, but those competitors could provide shade and some relief from water stress. Adapting to disturbances such as fire and outbreaks of insects also involves tradeoffs; we will see these in Chapter 6.

A complex tradeoff that is important in dense forests is associated with *shade tolerance* (how a species survives and grows when light is limited). Trees are often

---

[14] In the same sense, as we learn in high-school physics, as for gravitational potential energy or electrical potential (voltage). When this energy is catalyzed and released, it becomes useful energy, for growth (biological), power (electrical), or colliding bodies (gravitational).

[15] Because of this, the paleo-climate reconstructions that I discussed in Chapter 3 require careful selection of trees for sampling tree rings. Trees experiencing minimal competition, often at the extremes of their ranges, are considered optimal.

[16] Recall our discussion of leaf-area index (LAI) in Chapter 2 (Note 12). The term "closed-canopy" is used loosely, and not always consistently, to indicate LAI > 1.

[17] For example, the *gymnosperms* (our conifers dominant in the West) probably originated in the late Carboniferous Period, around 320 million years ago. *Angiosperms* are probably more recent, but there is some disagreement about their origin, putting it somewhere between 250 and 200 million years ago.

classified simply as shade-tolerant or shade-intolerant, with perhaps a middle category or two, but more precise *metrics* (standard numerical measures of something of interest) are possible.[18] In explaining why species persist and grow where they do, we often see what seems to be an evolutionary "choice" between shade tolerance and drought tolerance. A case can be made for a reasonably clear *gradient* between the two. I will suggest below that there may be clear winners and losers along this gradient in a warming West.

Big trees might seem intuitively as if they should be impervious to something like a 1°–2° increase in average global annual temperature. To some extent that is true. In the West we have trees as old as 4000 years, and many of the giant sequoias in the Sierra Nevada and different conifers in the Pacific Northwest are over 1000 years old (Fig. 5.2). Their lifetimes have included substantial swings in climate, though granted not so severe as what we are expecting later in this century under the "business as usual" socioeconomic scenarios. These trees have also weathered extreme winters and summers, certainly outside the averages[19] projected for the next few decades (Fig. 5.3).

Seedlings are much more vulnerable. We know from both years of observations around the West, and from both field-based and more controlled experiments, that the tolerance of seedlings for extremes of temperature and precipitation, both acute and seasonal, is much lower than that of mature trees. They are most vulnerable in the first year of life (Fig. 5.4). It is therefore inevitable that wins and losses in the aforementioned evolutionary game will come more quickly in an environment (forest) of seedlings than one of mature trees. Mature trees, particularly those of long-lived species, act as a buffer for forests in the face of a rapidly warming climate, but only up to a point, what we call a *lagged* effect.[20] The near future of Western Mountain forests thus depends crucially not only on the baseline changes in climate, but also on factors that determine the forests' *age structure* (what proportions, and which species, are seedlings, saplings,[21] and mature trees), with younger forests and those with short-lived species being vulnerable to more rapid or severe changes.

---

[18] For example, a *photosynthesis light-response curve* is a graph with incoming light on the X-axis and a plant's photosynthetic rate on the Y-axis. These graphs usually start up steeply and level off, under "light saturation". A curve that starts the most steeply, but levels off quickly, is probably a *shade-tolerant* species. It doesn't take much light for it to assimilate at its maximum rate.

[19] But probably well inside the projected extremes. More on this topic in Chapter 8.

[20] For example, the dominant trees in many forests across the West established in the *Little Ice Age* (LIA), roughly 1300–1850, depending on whom you ask, and with particularly cold periods around 1650, 1770, and 1850 (but which probably hit Europe harder than the American West). If ours were an equilibrium climate that had lasted a thousand years, instead of considerably warmer than the LIA, our "undisturbed" wilderness forests might look quite different.

[21] I will use this term as a rough equivalent of "adolescent". Numerous more precise definitions, for various purposes, can be found in forestry.

Fig. 5.2 Giant sequoias in a stand with other species at 7000 ft. elevation, in a cold-air drainage in the Mariposa Grove at Yosemite National Park. Photo by the author

Fig. 5.3 Foxtail pines, likely over 1000 years old, at 10,000 ft. elevation on the eastern slopes of the Sierra Nevada. Photo by the author

Fig. 5.4 Whitebark pine seedling at high elevation in the Cascade Range. The log may have provided just enough protection for it to survive. Photo by Alina Cansler

Forest Succession

Some tree species in the Western Mountains are indeed long-lived, such as the conifers mentioned above. Nevertheless, their lifespans, measured on *ecological time scales*, are shorter by *orders of magnitude* than the *evolutionary*[22] and *geological* time scales that I referred to in Chapter 1. So individual trees are replaced more quickly than they evolve, and the changes we can observe in response to a warming climate are changes in species composition rather than species adaptation in the evolutionary sense.[23]

We call *succession* the process by which forests, and plant communities in general, change over time in composition and structure. Succession will occur even in a completely stable climate. It is a process that is *emergent*[24] (an observable out-

---

[22] Remembering, of course, that evolutionary time is itself many orders of magnitude different for some organisms than for others. For example, *evolutionary* time scales for bacteria are shorter than *ecological* time scales for the long-lived trees that we are discussing. This scale comparison is limited to organisms with similar lifespans, specifically with respect to reproductive "turnover".

[23] But this is partly a consequence of the unprecedented speed of the current warming. Most climate changes in Earth's past, even the fairly rapid PETM (see Chapter 3), were more in sync with the evolutionary time scales of long-lived organisms.

[24] We will see this term later, in Chapter 8, as an element of complex ecological systems that can be difficult to predict even with detailed knowledge of fine-scale processes. That can be a challenge, because the *emergent properties* of systems are often the ones that we want to predict.

come of finer-scale interactions) from the interactions among species with different *life-history* traits. Some of these traits are shade tolerance and drought tolerance, growth and reproductive rates, and resistance to fire, insects, and pathogens.

In some stands in some forests, succession seems almost orderly, particularly after a *disturbance* (fire, landslide, volcano, windstorm) that is severe enough to kill most of the biomass aboveground. *Early-successional* species, not necessarily trees, will establish themselves first, but may not remain dominant. Life-history traits, such as the ability to extract nitrogen, an essential nutrient, from the atmosphere when it may be mostly absent in the soil in an accessible form,[25] can give these species an advantage early in succession that they will lose later. In the classic narrative for the West, our dominant conifer species establish later, when there is sufficient nitrogen in the soil in usable forms.[26] Once trees are dense enough to reduce light levels within the canopy, shade-tolerant species gain an advantage. The classical directional successional pathway moves through dominance by shade-intolerant species to dominance by shade-tolerant species.

This classical pathway is only one of three patterns that interest us in the contest of a warming climate in the Western Mountains. The second pathway is a simple one: the forest comprises one or two species, usually conifers,[27] that will remain dominant regardless of how old or dense the forest becomes. Ponderosa pine (or pinyon pine with juniper) form forests and woodlands in the Southwest and on dry eastern slopes of all the major ranges. Lodgepole pine forms dense forests in Yellowstone, other areas of the Rocky Mountains, and the Sierra Nevada and Cascade Ranges. These species are drought-tolerant, though not shade-tolerant, and also can thrive on volcanic soils, granting them sole "possession" (in the tree layer) of much of the western landscape.

The third pathway of succession is the most complex, involving multiple species and shifting patterns of dominance in both space and time. A mix of blind alleys, meandering country roads, and loops replace the one-way "freeway" of directional succession that is the best understood.[28] Commonly seen as having "multiple

---

[25] This process is called *nitrogen fixing*, and it will suffice to know that no one expects this to change much in a warming climate, at least not until extreme changes occur that are outside our time frame here.

[26] For conifers, "usable" generally means ammonium ions (NH_4^+); for broadleaf trees nitrates (NO_x^-). Nitrate pollution, whether or not associated with climate change, can certainly affect ecosystems, but is unlikely to affect this early stage of succession.

[27] An experienced traveler in the West might object that there are many stands of pure aspen in the Rocky Mountains and eastern slopes of the Cascade Range, and mixed stands of broadleaf trees in riparian areas. True. Species composition in the riparian may be stable, in that frequent disturbance (e.g., flooding) brings back the same species that dominate. The case of aspen is unique, and it is not yet understood (meaning that there is considerable disagreement) how stable aspen stands are in space; they may be a continually moving mosaic.

[28] Partly because it is more or less linear, but also because it is the paradigm that was long taught in forestry schools, in which timber management is a prime objective. The key element was that the most valuable timber species, for example Sitka spruce and Douglas-fir in the Pacific Northwest, dominated in middle succession, so "late-successional" became synonymous with "over-mature" or simply "decadent".

pathways",[29] forests of this type in the West are rich in co-dominant species (they may not have more total species than forests that follow the classical pathway, but their distribution is more even). With many species whose life-history traits differ by not so much (as opposed, for example, to two with widely different shade and drought tolerance), relative dominance is sensitive even to slight changes in the environment.

These slight changes can be external, such as climate, or internal, such as increased canopy density or small gaps from fallen trees. The subtleties and complexities of this topic are intriguing, and I encourage interested readers to use your web-search skills and track down some of the recent literature. For our purposes, however, I will state what may be an obvious focus. For the Western Mountains, we want to understand two external controls on succession: the effects of a change in limiting factors, from energy to water, and the effects of changing *disturbance regimes* (the overall character of disturbance patterns).

Dispersal: Keeping Up?

If climate is the "overarching" control on tree distributions, then trees will not survive where climate is unsuitable, but it should be clear that trees will not always be present where there is a suitable climate.[30] Species will not persist—they will lose in the evolutionary game—when they are out-competed for resources by other species for which the climate is also suitable, or even more suitable. Before that game is played, however, species must encounter that suitable climate. Because trees are *sessile* (organisms that don't move, in contrast to *vagile* organisms, i.e., most animals, which do), most of our Western Mountain conifers can only *disperse* [31] between generations when their seeds are moved.[32]

---

[29] A (really, almost prehistoric) classic paper laying out this paradigm, for one particular example of it, is P.J. Catellino *et al.* 1979. Predicting the multiple pathways of plant succession. *Environmental Management* 3:41–50. A more recent exposition is D.C. Donato *et al.* 2012. Multiple successional pathways and precocity in forest development: can some forests be born complex? *Journal of Vegetation Science* 23:576–584.

[30] For those of us who had elementary schooling in logic, and still have at least a dim memory of it, this is an example of the *converse* of a statement or theorem not necessarily being true even if the statement itself is true. Elementary, yes, but it can become complicated quickly, as can everything in ecology. A good way to avoid many pitfalls is to remember, as I said above (Note 8), that "absence of evidence is not evidence of absence".

[31] Here we should distinguish among *dispersal, migration,* and "*translocation*", as they are used technically. Dispersal is a permanent move, either by seeds (plants) or juvenile or adult animals away from the place of their birth. Migration is a seasonal move, followed by a move "back" within the year (the unfortunate phrase "assisted migration", though widely used, really refers to assisted dispersal or translocation). Translocation, or assisted colonization, or "managed relocation" refer generally to human-assisted moves.

[32] A few conifer species can "move" by clonal expansion, as can many broadleaf species. This will certainly affect local species composition, as anyone who has watched the expansion of a clone of quaking aspen can report, but it is clearly much slower than seed or juvenile-animal dispersal.

The two main *vectors* (jargon for the means by which something happens) for seed dispersal for conifers of the West are wind and animals: mammals and birds. Wind is more important for sheer volume, because most of our conifer species' seeds have *wings* (literally) that are up to four times as long as the seeds themselves. Animals deposit seeds either intentionally or unconsciously (eating and later defecating). An example of intentional dispersal is that squirrels and mice *cache* seeds for future use; so do birds.[33]

When we try to estimate where seeds will disperse, we come up against an obvious challenge: you can't always predict how hard and in which direction the wind blows, or how far and in which direction an animal will move. Most seeds won't go very far, but some could go a long way given the right conditions, such as a strong persistent wind.[34] Mathematical models, called *dispersal kernels*, provide estimates of dispersal distance that are *probabilistic*: instead of calculating a single number they estimate a distribution of distances. We saw earlier how a bell-shaped curve, the *Gaussian* or "normal" distribution, is a mainstay of calculating the uncertainty (in our scientific sense) of many statistical estimates (see Fig. 1.2). The normal distribution is symmetric, with the left side a mirror image of the right. Dispersal distributions are different, because most distances are short, with a few long ones, so these are called *right-skewed*. Some right-skewed distributions are well behaved and usable, whereas others, called *fat-tailed* (basically a lot more likelihood of big game-changing events), are difficult to use. We shall refer to more fat-tailed distributions in Chapter 8, regarding extreme events. Dispersal distributions are generally assumed to be of the manageable (well behaved) type, and so (getting back to our purpose) we have some confidence that we can predict average rates at which tree species might disperse.[35]

What we need to know for each species is whether its dispersal rate is fast enough to "keep up" with an environment that is changing with a changing climate. At the broadest scale, there is a warming "movement" of climate along a latitudinal *gradient* from south to north. For example, we expect that the average annual temperature at 45° N latitude today will be the average at 48° N at some time in the future. The speed of this movement is often referred to as *climate velocity*: average temperatures "moving" north. If all other things, such as elevation, precipitation, and

---

[33] For example, as we will see in Chapter 7, the Clark's nutcracker caches seeds of whitebark and limber pine, then often "forgets" them, to the advantage of hungry grizzly bears in autumn.

[34] In theory, with the right timing, a migrating bird could transport a seed for hundreds of miles, depending on the speed of both its flight and its digestion. For example, Arctic terns have been known to migrate tens of thousands of miles in a year, but birds that cache seeds (i.e., can hold on to them) do so within their local home ranges.

[35] Whether or not the climate is changing. That doesn't really affect dispersal distances directly in any way that we can measure, but as I discuss next, it affects whether dispersal is successful.

seasonal patterns, were equal,[36] the optimal environment for each species would move north (or south in the Southern Hemisphere) at the climate velocity.

With no barriers to dispersal, it is likely that our conifer species in the Western Mountains could keep pace with climate velocity.[37] On real landscapes, when we try to estimate the potential for dispersal, things become complicated quickly. For example, let's consider a relatively straightforward case. Recall our discussion of the Sky Island ecosystems in the Southwest in Chapter 2. Each is an isolated forested, and sometimes alpine, ecosystem in a *matrix* (landscape ecology jargon for the surroundings of a particular area of interest) of lower-elevation shrubland or desert. For any species, plant or animal, to disperse from one Sky Island to another with more favorable climate would require navigating that hostile terrain.[38]

Barriers within mountain ranges can be just as formidable as those between them. For example, favorable winds between Sky Islands (southerly, in this case) could easily provide better transport for seeds than the complex variable winds within a range, which are subject to fine-scale constraints. For trees, the hostile matrix of shrubland and desert (too hot and dry for seedlings to survive) means an all-or-nothing situation for dispersal. Wind must transport seeds all the way to the next Sky Island. But the matrix within a mountain range can be just as hostile, for example, a glacier or permanent snowfield, or a rocky ridge or series of ridges.

When we try to estimate the probability that a tree species will disperse successfully in a warming climate, the complexities mount quickly. We may try (and yes, some have attempted this) to identify a "path" north across a mountainous landscape for a species of interest, but we run up against the problem of *cumulative error*. This refers to the compounding of uncertainties in a series of calculations, such as calculating the probability of dispersing by wind over a series of ridges consecutively. Before too long, we may be "off" by at least the average distance between two of those ridges,[39] or if asked for a yes or no answer, be unable to say either with any confidence.

---

[36] And of course they will never be equal. By definition, elevation changes quickly and continuously in mountains. Average *lapse rates* (relatively consistent changes in temperature with elevation) can be estimated, but they can vary widely, and with elevation-dependent warming (see Chapter 3) they may lessen. Topographic influences on precipitation also vary widely, meaning that the interplay of energy and water limitations is harder to estimate than if we could rely on broad averages.

[37] As far as I know, no one has tried to do these computations rigorously, for an obvious reason: barriers are virtually everywhere. This has not stopped people from creating climate velocity "grids", for example, 1-km resolution grids for North America based on output from GCMs. For example, see https://adaptwest.databasin.org/pages/adaptwest-velocitywna.

[38] Easiest for birds, more difficult for un-winged animals that have to walk across, and for plants, which have to ride the wind.

[39] For a toy example, suppose we have a series of ridges about a half-mile (800 m) apart. There is a 90% probability that a seed will disperse 800 m with average winds before the environment becomes too hostile. Consider the simple probability that a species will disperse successfully across 5 ridges (with our numbers $0.9^5 = 59\%$). What if our 90% were off by 5%? $0.85^5 = 44\%$. More likely than not easily becomes less likely. This doesn't mean that we shouldn't try, only that we should be careful with our claims.

The past can provide us some controls on our estimates, as is often the case. For trees, we have two main types of evidence that help us understand the limits to dispersal over time. The first is very qualitative, but instructive. We can look for where species might be, based on suitable habitat, and ask why they are not there. If the species is nearby, the answer may be that it could not compete, for some reason. If it is nowhere nearby, the answer is the lack of a means for dispersal. For example, two natives of California, coast redwood in the Coast Ranges and giant sequoia in Sierra Nevada, will thrive, if planted, in the lower elevations of the Cascade Range in Washington State. Why are they not found there? Across 500–1000 miles of mountain terrain, the number and strength of the barriers to dispersal were enough that even 18,000 years (since the retreat of the ice sheets from western North America) was not enough time to surmount them.[40]

The paleological record also reveals limits to dispersal over time. Because fossil pollen and macrofossils can be dated with decadal accuracy, and climate reconstructions from tree rings and other sources to as fine as annual accuracy,[41] we can align changes in the locations and abundance of the plant fossils with changes in climate. When the climate shows persistent trends, beyond interannual variation, we can compare estimates of past climate velocities with estimates of the average rate of dispersal of the plant fossils. We get a *lower bound* [42] on species' average potential dispersal rate. If we make the reasonable assumption that barriers to dispersal, at least topographic ones, are constant in *ecological time*, then we can *calibrate* (put some realistic bounds on a process based on empirical evidence) our projections of which species might keep up with climate velocity, and where they might do so.

Patterns of the past serve us only so well for projecting future dispersal success. As I said in Chapter 3, the current rate of climate warming is almost certainly faster than any previous one, and definitely faster than any in the plant-fossil record. This means that we are in new territory for climate velocity, and we have no guarantees that the associations between it and dispersal are even as well behaved as those in the past may seem to have been.[43] New limiting factors may arise; for example, topographic barriers may make dispersal at the new necessary rates so improbable as to be doomed to fail. If we want species to find new suitable habitats, we may have to move them.[44]

---

[40] One could argue that this is a bit too simple. It could be that there were enough dispersal "events" of a certain distance to enable this movement, but that seedlings were always out-competed by other species. An implicit assumption of climate-velocity models is that the "landing zones" for dispersing species will give them the same competitive advantage in the future as they have now in their current habitat.

[41] In a situation like this it is best to "blur" annual estimates of climatic variables, to avoid false precision.

[42] Why a lower bound? The species was at two places at time 1 and time 2, so it could move at least fast enough to span the distance. But remembering that "absence of evidence is not evidence of absence", it might have dispersed farther, but we haven't observed that.

[43] This is another instance of the problem of *non-stationarity*, to which I have referred before and will again.

[44] We wrote a paper about this. L. Iverson and D. McKenzie. 2013. Tree-species range shifts in a changing climate—detecting, modeling, assisting. *Landscape Ecology* 28:879–889.

Carbon: Source or Sink?

So far I have written as if forests are all about trees. Sure, they are the *sine qua non* of a forest, but they are only a part of a forest's budget of one of the major "currencies" of global climate change, which is carbon. As we saw in Chapter 3, our understanding of the strong association between atmospheric CO_2 and global average temperatures is based on the paleological record of the global carbon cycle,[45] of which CO_2 is part but not all. CO_2 is emitted into the atmosphere from respiration, biomass burning, and other sources, and extracted by plants and physical or chemical processes such as *weathering* (the breakdown of rocks and other minerals that draws carbon into the ground). The balance of emission and extraction controls levels of atmospheric CO_2.

The role of forests in the global carbon cycle is complex, with subtleties[46] that could take us far afield. I shall be content to discuss the balances at a fairly crude level. What concerns us is whether forests are a net *source* (emitting more carbon than they extract) or a *sink* (the opposite), which forests are which, and whether that might change with ongoing warming of the climate. To do that we need to be familiar with where carbon is, how vulnerable it is to disturbance (more in Chapter 6), and more specifically how quickly carbon is gained or lost.

Plant biomass in forests (and elsewhere, of course) is both living and dead, both woody and non-woody, and above ground and below ground. Above ground, woody live biomass includes *boles* (trunks) and branches of trees and stems and branches of shrubs. Non-woody live biomass includes foliage of all types and the stems of grasses and herbaceous plants. Below ground, woody live biomass includes roots of both trees and shrubs. Non-woody live biomass includes various tissues that are part of roots.[47] Woody dead biomass above ground includes boles and branches of dead trees and shrubs, whether still standing or fallen. Below ground it is dead roots. Non-woody dead biomass above ground is dead foliage, whether still on a plant or on the forest floor. Below ground it is the dead tissues that are part of roots. On average, roughly, about half of forest biomass is carbon, by weight[48] (*mass*).

---

[45] We all know that carbon comes in other forms besides CO_2, but CO_2 is the principal "vehicle" by which carbon enters and leaves ecosystems. I use it and "carbon" somewhat interchangeably.

[46] For example, a useful metric is the *Net Ecosystem Exchange* (NEE), the gain or loss per unit time of CO_2 from an ecosystem. This can be measured in "bulk" by various sensors, thereby characterizing a forest as a net source or sink. On regional to global scales, it can be used to inform *Earth-system models*, which incorporate exchanges between the atmosphere and the land surface (and the oceans) to project response to climate change.

[47] Even though these latter are a tiny fraction of the total forest biomass, their biology can be important for the health and resilience of forests. See Note 13.

[48] A reminder that *weight* is actually a measure of force, not the English-unit analog of the metric units of mass. A person of a certain mass will have that same mass on the moon, but only about a sixth of the weight; weight = mass × the local *acceleration* due to gravity, in the right units, whether English or metric. For those who remember their elementary physics, $F = ma$.

Biomass does not account for all the carbon in forests. Huge reservoirs of carbon are in soils, particularly those undisturbed by humans, whether partly by agriculture or permanently by urbanization. For example, as I write, probably about 3.5 times as much carbon is in the Earth's soils as in its atmosphere; this is at least two orders of magnitude higher than annual global fossil fuel emissions. About a third of soil carbon is *non-organic* (not biomass), in carbonates and other minerals. In most intact forests, soil carbon changes much more slowly than live biomass above ground, on a scale of decades or even centuries as opposed to years. All forest soils are vulnerable to disturbances such as fire, however, and some may be near *thresholds* of warming temperatures beyond which soil carbon could be released rapidly.[49] For example, soils in peatlands in the sub-polar regions are considered particularly vulnerable.

How do we proceed to understand how the *source-sink dynamics* of forests (how much carbon they extract vs. emit) will change in a warming climate, especially in the Western Mountains? Let us return to our concept of *limiting factors*, especially energy and water, remembering that when a limiting factor lessens, whatever it limits (here, growth) increases. In an energy-limited environment, increasing temperatures and CO_2 will reduce energy limitation (remember that CO_2 provides energy in the form of nutrients via photosynthesis). In a water-limited environment, increasing temperatures will exacerbate water limitation. CO_2 may increase *water-use efficiency* (literally, plants don't need as much water to accomplish their biological tasks), partly offsetting the increased water limitation from higher temperatures.[50] From these fairly simple notions we can infer that in general, a warming climate in the West will move existing forest ecosystems away from energy limitation and toward water limitation. Depending on where they are on this *gradient*, they will also change position on a gradient between *source* and *sink*, all other things being equal.[51]

Imagine now a map of the West, with the mountain ranges already delineated as in Fig. 1.1, and now imagine overlaying a patchwork of two colors—green for energy limitation and brown for water limitation. Assuming that we had such a map, showing the spatial patterns of energy and water limitation, we might expect that in a warming climate some green patches will change to brown (and darker greens to

---

[49] More of this below in "Feedbacks to climate change" and in Chapter 8.

[50] The evidence for this is from small-scale controlled experiments. The complex task of extrapolating these experiments to the larger-scale dynamics of forests is still a challenge.

[51] Which of course they are not, but bear with me for a bit. We will see that the energy-water gradient forms a robust baseline for variation introduced by both environmental pressures and human activities.

lighter, but lighter browns to darker).[52] At a finer scale, imagine next how the limiting factors change within one patch during *forest succession*. Over the transition between a young open-canopy forest and an older closed-canopy one, some trees are shaded out, thereby becoming more energy-limited, while also competing against more others for water, thereby becoming more water-limited. Within a larger-scale trend from green to brown, we would see considerable variation and patchiness. Now imagine the change in each tree's ability to acquire and use resources as it grows larger: more foliage means more raw material for photosynthesis, but more biomass overall means more overhead for maintenance of what is already there. At the finest scale, energy and water limitation still vary. The multiscale picture shows a trend at the coarse scale, but (mostly) variation at finer scales.

All these considerations, and many more details, create challenges for projecting the fate of forests in a warming climate, not just the source-sink dynamics, but also forest succession and the possibility of drastic changes in the extent and composition of forests (see "Forests on the brink", below). It is easy to get lost (or obsessed with, if a researcher) in the details. They change, as do the *parameters* in the models that we build of them, but what persist are the limiting factors, and we will continue to use them as a guide through the complexity.

Feedbacks to Climate Change

Whether or not forests are sources or sinks of carbon clearly has a direct effect on the *climate forcing* (see Chapter 3), by adding CO_2 to the atmosphere or removing it. Before we sum up[53] the effects of forests on global warming, however, we have to consider feedbacks: whether positive or negative, how strong they are, and whether they are global or regional to local. The feedbacks discussed in Chapter 3, relevant on the global scale, are also important links between forests and the atmosphere.[54]

---

[52] I have shied away from drawing such a map, because it would be fake while tempting viewers, including me, to view it as real. This is something maps are known to do. We are nowhere close to having such a map, but here are some things that might happen in a warming climate. (1) Cold high-elevation forests that were deep green might become less so, (2) arid low-elevation forests that were brown (of all shades) would become more so, (3) forests near the center of the gradient, on the green side, could turn brown, (4) forests shifting toward the center would shift toward the "sink" end of the associated source-sink gradient, whereas those shifting away from the center would shift toward "source". Do you see why #4 is true, from our discussion of limiting factors? And knowing what the forcings are, do you also see why practically all movement would be away from the green end and toward the brown end? Another thing that would happen is that the pieces of the patchwork would change shape and size.

[53] And I won't do this explicitly, with numbers. Accounting methods vary, as do the numbers from year to year, and the *stationarity* of the calculations, whose *parameters* may change.

[54] For a not too technical summary of forest feedbacks to climate change, see G. Bonan. 2008. Forests and climate change: forcings, feedbacks, and the climate benefits of forests. *Science* 320:1444–1449.

The *surface albedo* of forests is lower (i.e., reflects less short-wave radiation) than that for other types of land cover. So when forests are cleared, by human or natural disturbances, albedo increases, creating a negative feedback (cooling) that offsets a warming climate. When forests grow back, or are planted, albedo decreases, creating a positive feedback, accelerating warming. In areas with perennial or even seasonal snow cover, the difference between non-forest and forest is greater, because few types of land cover have higher albedo than snow. Conifer forests, such as those that make up most of the forests in the Western Mountains, have even (slightly) lower albedo than deciduous forests.[55]

Cooling from *evapotranspiration* (ET: see Chapter 4) is a negative feedback to climate warming. Because ET depends on the total surface area of plants, there is more ET from forests than from other land covers. Recall from our discussion of water-balance deficit (DEF) that the realized, or "actual" evapotranspiration (AET) is usually less than the potential (PET), depending on whether sufficient water is available for plants to transpire "fully". The negative feedback is therefore less if forests are water-limited than if they are energy-limited.

The cooling from ET is not entirely free, however. It releases water vapor into the atmosphere, and that is a potent greenhouse gas,[56] though not so long-lived as CO_2, and not well mixed, so its effects are more local. It turns out that in some regions, notably the forested areas of the Arctic, the (short-lived warming) climate forcing from added water vapor (from ET) more or less cancels out the cooling effect.[57]

Our last feedback is an interaction between *source-sink* dynamics and albedo. If CO_2 fertilization changes forest albedo significantly, i.e., decreasing it by increasing the coverage of the land surface by needles, then the ensuing drop in reflectance warms the atmosphere more. But this may be offset by increased ET from plant surface area.[58] As I write, I know of no one who has tried to run numbers on this interaction—perhaps you can see why.

Globally there is no question that the more forest there is, the more carbon is sequestered that might otherwise be in the atmosphere as CO_2. How to maximize forest cover, with the associated uncertainties and risks, is a topic I will visit briefly in Chapter 9, but with our focus on the Western Mountains, which of these

---

[55] For perspective, the albedo of snow is about 0.8 (out of 0.0, the lowest possible, and 1.0, the highest). Deciduous forests are between 0.15 and 0.18; conifer forests between 0.08 and 0.15, about the same as worn asphalt; open ocean (without sea ice, which is 0.5–0.7) is 0.06. Green grass is about 0.25. So forest vs. non-forest matters, but snow vs. non-snow matters a lot more.

[56] More to the point, there is a positive feedback between water vapor and CO_2 that amplifies global warming, because a warmer atmosphere can hold more water vapor.

[57] For the particulars, see A.L.S. Swann *et al.* 2010. Changes in Arctic vegetation amplify high-latitude warming through the greenhouse effect. *Proceedings of the National Academy of Sciences, USA.* 107:1295–1300.

[58] Where will these things happen? If you recall our discussion of leaf-area index (LAI: Chapter 2, Note 12), with LAI above 1 the albedo is not likely to change much, whereas moving from 0.3 to 0.8, for example, will clearly reduce it. In general, this feedback will be more relevant in arid mountain forests with less than 100% canopy cover. We have a lot of these in the Western Mountains.

feedbacks will dominate, and where? There is some consensus that in the *boreal* (high-latitude) forests, principally in Canada, Alaska, Siberia, and Scandinavia, the positive (warming) feedbacks will dominate. More forest cover = lower albedo, not compensated for entirely by greater ET. For tropical forests, what consensus there is points to negative feedbacks: the cooling effect of ET exceeds the warming effect of lower albedo. For temperate forests, like all ours in the West, there is enough uncertainty about the numbers, and things are variable enough, that the question is still open. Perhaps surprisingly, though, we can make some observations about what specifically might happen where.[59]

In a *model experiment* (running a simulation model with *treatments* and *controls* like a field or lab experiment), Swann and colleagues asked the question "What would happen to regional climate across the USA if all the forest in a particular region were lost?" They divided the continental US (CONUS) and Alaska into 18 regions and simulated the future climate with an *Earth-system model* (basically a GCM with our feedbacks, and other inputs from the land surface), artificially removing all the forest from each of the 18 regions in turn. What they found was that forest loss in some places matters a lot more than in others, and it is not always the local climate that is most affected by forest loss. Changes in the feedbacks are strong enough to affect large-scale patterns of circulation, which are the drivers of regional climate. For example, forest loss in the southern Rocky Mountains raises summer temperatures most not only in the northern Rocky Mountains (perhaps not too surprising) but also in the Pacific Northwest. In contrast, forest loss in the "Pacific Southwest" (including the Sierra Nevada and southern parts of the Pacific Coast Ranges) has little effect across the West, but does raise temperatures east of the Mississippi River. These non-local effects (*teleconnections*) are not intuitive and often complicate our studies of the effects of climate change.

Forests on the Brink? An Example of Detection and Attribution

A global loss of forests, like rising sea levels, is a major environmental concern in a warming world. Rising sea levels are an inevitable consequence due simply to one of our "constants" from Chapter 1: the melting (freezing) point of ice (water). Forest loss may be equally inevitable, though not so simply explained. Worldwide there is far more acreage below treelines than there is above. When a suitable environment (for trees) moves up in elevation because average global temperatures are increasing, the land available to support forests shrinks, and even in the ideal case where forests could move instantly there is forest loss.

---

[59] This discussion is based on A.L.S. Swann *et al.* 2018. Continental-scale consequences of tree die-offs in North America: identifying where forest loss matters most. *Environmental Research Letters* 13: (055014—paper identifier).

Sea-level rise from a warming global climate is predictable from the loss of the continental ice sheets of Greenland and Antarctica,[60] because we know that when that ice melts it winds up in the oceans. Forest loss is also broadly inevitable, but not so predictable, because no such constants (like 32 °F) obtain. But forest loss is observable, like sea-level rise.[61] The *detection* of forest loss, sometimes called "dieback", is already happening in many places around the world, with trees dying *en masse* fairly quickly.[62] But can we *attribute* this widespread mortality to global warming? More precisely, is it the consequence, either directly or indirectly, of a relatively small (1 °C) increase in the global average temperature or in its more regional variations, such as *elevation-dependent warming* (EDW) or *latitude-dependent* (more in polar regions) warming.[63]

In the Western Mountains, a warming climate could remove trees in three ways. The direct way is by simply becoming too hot (and usually dry), rapidly crossing some physiological threshold.[64] The two indirect ways are (1) some *agent* of mortality attacks trees that are weakened by heat, or drought, or both (Fig. 5.5), and (2) a high-severity disturbance, such as wildfire (likely more widespread in a warming West—see Chapter 6), kills most or all of the trees at some location and seedlings cannot survive where mature trees could (Fig. 5.6). Most of the cases of either dieback or regeneration failure in the West are one of the indirect types, so a question arises. "Don't these happen even in a stable but variable climate?" Yes, but the frequency and magnitude of these events may be increasing, which is a strong indication, if true, that we can attribute them to global warming. In keeping with scientific caution, as I write the researchers in this area are circumspect about the *attribution*, but clear about the *detection*, and they are on high alert about the consequences of increasing world-wide forest dieback.

---

[60] The informed reader will notice that I am ignoring another contributor to sea-level rising: warming ocean temperatures. For purposes of comparison, I am also ignoring the loss of glaciers and perennial snow outside of Greenland and Antarctica, which contribute a bit to sea levels.

[61] And much more evident without accurate instruments. Whereas sea levels change in increments of millimeters, forests change in increments of squared kilometers (six orders of magnitude difference).

[62] A fairly recent review of observations, with some hypotheses about causes and uncertainties, is C.D. Allen *et al.*, 2010. A global overview of drought and heat-induced tree mortality reveals emerging climate change risks for forests. *Forest Ecology and Management* 259:660–684.

[63] Recall from Chapter 3 that the 1° increase is our best evidence of a warming climate *per se*. Other observations, such heat waves, are considered to be the effects of the global rise, *via* feedbacks and changes in circulation.

[64] For a detailed discussion of such thresholds, see N. McDowell *et al.*, 2008. Mechanisms of plant survival and mortality during drought: why do some plants survive while others succumb to drought? *New Phytologist* 178:719–739.

Fig. 5.5 Dead trees of multiple species in 2015 on a southwest-facing (hottest and driest) slope in Sequoia National Park, victims of a multi-year drought. Photo by Nate Stephenson

Fig. 5.6 A landscape in Yosemite National Park at 5000-ft. elevation that burned severely and in which trees may not regenerate, being replaced by shrubs. Photo by the author

Forest Models: How Can We Use Them for Climate-Change Projections?

There are probably more forest models than GCMs and eco-hydrological models combined. Long before climate change was on the radar of any but the most pre-scient researchers,[65] foresters and forest ecologists wanted to project conditions of forests either if succession was allowed to proceed without intervention or if they were managed. Unlike eco-hydrological models, in which trees are essentially green carbon, what I will call forest models simulate trees explicitly, either in bulk or as individuals. Like eco-hydrological models, however, the scale of interest controls their design. Forest models have many similarities to eco-hydrological models, so the discussion below draws heavily on Chapter 4.

Not all forest models can address questions about climate change. To do so, some aspect of climate itself must be a *driver* (basically a *variable* that has consequences for other variables). The driving can be either empirical, as with a predicting vari-able in a statistical relationship, or more biologically based, such as a *limiting factor* in plant physiology.[66] Empirical models of forests' response to climate change are *inverse models*. They seek contingent parameters (see Chapter 4) that populate sta-tistical relationships between metrics associated with climate and those associated with trees. Climate-based metrics are temperature and precipitation and many oth-ers derived from these; tree-based metrics are basic occurrence (are they there or not?), height and diameter growth, and probability of mortality. These empirical models are often called "bioclimatic envelope models".

If drivers are biologically based, forest models are often dubbed "process-based", meaning that a change in one or more biological processes is simulated explicitly. These are *forward models* (see Chapter 4), inferring the consequences of an earlier (in time) state for a later state, and are the tree-specific analogs of the GCMs and eco-hydrological models.

Process-based and bioclimatic envelope models answer different questions. As the analog to the models we have seen earlier, process-based models answer "what if…?" For example, we could ask "IF a particular future climate (from a GCM or regional climate model) were to be realized, what would be the expected changes in tree-species composition[67] for our watershed or region?" More prudently, we would add "What is the range of possibilities?", because we are compounding the uncer-tainties associated with climate and those associated with the trees' response.

---

[65] A personal observation: the first timber management models began to proliferate about when the sweep of timber extraction across the US had more or less reached the Pacific Ocean. Interest in models rises in a world with limits.

[66] Recall though from our earlier discussions of models that this distinction is not so clear as it may appear to (or be wished for by) some. For example, "biologically based" here is the analog of "physically based" in climate models, but often the *parameters* in the mathematical representa-tions of relationships are obtained empirically, probably through statistical estimation.

[67] Or we might even ask "What are the chances of forest dieback?" With the current imperfect state of forest models, we would be hard pressed to get that right.

The original intent for process-based forest models was to apply them at very fine scales, from one to a few forest stands. Climate was not always even a major driver, because modelers were most interested in ecological processes such as light or nutrient limitation from competition. With the growing urgency over the years of projecting the fate of forests at broad scales in the face of climate change, the scale of interest and the importance of climate both grew. Typically now we see forest models applied regionally and even globally; they are now often called "dynamic global vegetation models" (DGVMs), and almost always include explicit input from climate models. For example, a well known DGVM is run for the continental US at about 1-km[68] *grid spacing* (the size of simulated cells within which processes are modeled explicitly), with input from GCMs.

Bioclimatic envelope models ask "what is…?" instead of "what if…?" Their aim is to establish numerical relationships between (presumed) drivers and responses. If the drivers are climate, the implication is that when the climate changes, so will the response, in a predictable way. More specifically, if the new climate in location A is the same as the old in location B, then the same species that thrived before at B will now thrive at A.

The same pitfalls of modeling that we encountered in Chapter 4 obtain for forest models of both types. They are sensitive to initial conditions,[69] they are likely to change when applied at different scales, and they assume stationarity. We saw there how the pitfalls applied to process-based models (e.g., GCMs and eco-hydrological models). For empirical models, a key pitfall is the assumption of stationarity. The *parameters* in statistical relationships, which are the only thing that matters in these models,[70] must stay the same for the model to be valid. Because so many things are not accounted for in a model, this is unlikely. That said, in practice empirical models sometimes outdo their "process-based" counterparts. For example, if a bioclimatic envelope model predicts that a species will NOT be at a location, we are unlikely to find it there unless it was planted and then sheltered. But process-based models can be wildly wrong about locating a tree species—thousands of feet in elevation above or below where it really is—when a model is applied outside the region in which it was developed.

The two types of models are fragile in different ways. A reasonable strategy, though not often pursued, is to combine the knowledge gained from each. For example, empirical models excel at identifying parameters of the sort that populate the algorithms that drive process-based models. The large databases (i.e., sample sizes)

---

[68] More precisely at 30 arc s, about 0.924 km^2 at the Equator, but changing moving poleward. See D. Bachelet *et al.*, 2018. Translating MC2 DGVM results into ecosystem services for climate change mitigation and adaptation. *Climate*. Free online at https://www.mdpi.com/2225-1154/6/1/1/htm.

[69] For empirical models these are the values of the driving variables, since there is no evolution from one state to the next, by definition.

[70] Simplifying again. We solve for *parameters*, but also for the uncertainty associated with them. Even if those parameters are "right", the range of possibilities may be too large to give a meaningful answer.

usually associated with empirical models help to increase the *precision* (technically here, how much the variability can be narrowed) of estimates. Keeping in mind the stationarity pitfall, we can apply "what is", with precise parameters, to ask "what if" in a process-based model. As we saw in Chapter 4, the complexity of models is both friend and foe; it allows us to manage a welter of combinations in our "what if" questions that would otherwise be intractable, but uncertainties and errors are cumulative.

What Do We Expect for the Western Mountains?

We know that forests of the Western Mountains range from almost purely energy-limited to almost purely water-limited. We also have inferred that in general, movement induced by climate change on the gradient of energy-water will be toward water limitation. There have been many forest-model experiments for specific areas across the West, and they will continue. Some themes have emerged, but there have also been anomalies and contradictions. Rather than trying to summarize, classify, and evaluate all the outcomes, let us do an exercise in inference[71] from what we know. In general, western-slope forests (i.e., not in the rain shadow) should be less vulnerable to climate-induced stresses. In most areas, the competitive stature of more drought-tolerant species will increase.

The Cascade Range and the Pacific Coast Ranges (to the Klamath and Trinity Mountains)

The northernmost parts of these ranges, west of the crests, are the most energy-limited forests of the West.[72] Overall, because of the maritime influence and the regional projections (less than average warming) for the Pacific Northwest, we can expect these west-side forests to be the most *complacent* (jargon for "not affected that much") to warming, with some release of energy limitation at higher elevations where the energy limits are climate-based rather than nutrient-based.[73] East-side forests will be hit harder, especially at lower elevations, as they are already water-limited. Lower treelines may rise.

---

[71] And as we proceed through chapters of the book, I trust that the reader will bear with me as I add specific inferences incrementally to these sections based on the topics in each chapter.

[72] Geographically speaking. Of course, as we have seen, things also change with elevation, particularly on the drier aspects.

[73] Recall that nutrients limited by competition are an energy source.

The Northern Rocky Mountains

Climate is more continental here, and more warming is projected, on average, than for the Pacific Northwest. Even so, the productive forests of the western slopes share the other's energy limitation. We can expect some climate-based release of this at higher elevations, where there is sufficient input of water. The eastern slopes will be vulnerable in the same way as others.

The Sierra Nevada

This arid range could be hard hit by warmer temperatures accompanied by periodic or persistent drought. There is clear water limitation even on the western slopes, made worse by recent drought, and we can probably expect lower treelines to rise more than they might in the more northern ranges, particularly on the west side.

The Pacific Coast Ranges (South of the Klamath and Trinity Mountains)

These ranges are drier than their northern counterparts, and will be subject to the same (new?) circulation patterns as the Sierra Nevada. Water limitation is almost everywhere, so we can expect that to worsen. Forests that already have the most drought-tolerant species, such as pines and oaks, may lose a lot of trees to climate change.

The Southern and Central Rocky Mountains

The complex pattern of landforms in this region (see Chapter 2) belies a simple division into western and eastern slopes, although there certainly are prominent north-south divides mixed in throughout. Western slopes receive air from the Great Basin, with most of its moisture already gone, rather than from the Pacific Ocean. Like the Sierra Nevada, these ranges are mainly water-limited, even west-facing high-elevation forests (with a few exceptions). Once again, we can expect lower treelines to rise.

The Sky Islands and the Basin Ranges

These ranges, already marginally populated by trees at only the higher elevations, will likely be the hardest hit. If lower treelines rise much, forests will disappear altogether. It doesn't help that this region is expected to suffer from more drought than the rest of the West.

Chapter 6
Ecological Disturbance

Some say the world will end in fire,
Some say in ice.
From what I've tasted of desire
I hold with those who favor fire.
—Robert Frost

Wildfire in the West has been front-page news through the second decade of the twenty-first century, because of its effects on people, structures, landscapes, and the skies. Fire has been part of the Earth system for as long as there have been oxygen and fuel, however, probably since about 400–500 million years ago.[1] Since then, fire's role in ecosystems has ranged from very little or none, where either too cold or wet or too hot or dry,[2] to very important in most temperate ecosystems, boreal forests, and the dry (but not too dry) tropics. In the Western Mountains, fire is a part of most ecosystems.[3] It is one type of *disturbance* therein, and one of the most noticeable, although it is not the only one that can affect large areas. B*iotic* (caused by something living) disturbances, particularly insect outbreaks, can also affect large areas. Besides fire, other notable *abiotic* disturbances are windstorms, avalanches, landslides, and earthquakes. Though covering less area individually, landslides and earthquakes are frequent enough in some mountains to affect more area than fire or insects.

---

[1] Early in Earth's history, there were of course volcanic eruptions and other fiery events. I am referring to the origins of wildfire as we know it, fueled by plant biomass and supported by enough oxygen in the atmosphere to sustain combustion, between about 16–30%, with the current level being 21%. Below 16% fire will not propagate, and levels above 30% are possible only with enough moisture in the air to extinguish any fire that might start even if it had plenty of oxygen.

[2] For example, some deserts, polar regions, and rainforests, but not all. Just a little fuel, and a dry spell, can be enough.

[3] We wrote a whole book about this, also published by Springer. D. McKenzie, C. Miller, and D.A. Falk. 2010. *The Landscape Ecology of Fire*. Springer, Dordrecht, The Netherlands.

© Springer Nature Switzerland AG 2020
D. McKenzie, *Mountains in the Greenhouse*,
https://doi.org/10.1007/978-3-030-42432-9_6

What Is a Disturbance?

Fire, insect outbreaks, and most land movements (*i.e*, avalanches and landslides) are all examples of what I will call *ecological disturbance* in this book. Readers who have more than dabbled in the ecological literature may have noticed considerable variation in what is called disturbance, and in definitions of it. For our purposes, an ecological disturbance is a "relatively discrete event in space and time" that "knocks an ecosystem off course by a bit or more".[4] A continuous phenomenon that does not change an ecosystem abruptly is therefore not a disturbance. By that reasoning, I exclude drought, because it is at least seasonal and can be decadal, background air and water pollution, and pressure from ongoing human use of national parks and wilderness, for example. But I also exclude here events that are obviously disturbances, by our definition, such as clearcut logging and earthquakes, if they are not directly affected by a warming climate.[5] For convenience, we will call disturbances *natural* if they have no sociological dimension (as does logging).

I also distinguish between disturbances that are or aren't *contagious*. A contagious disturbance needs a medium that supports its spread, along with the force that moves it.[6] The medium for both fire and insects is vegetation: fuel for either combustion or metabolism. I shall consider windstorms and land movements to be non-contagious, although contrary arguments have been made.[7] Our main focus is on ecological disturbances that current common wisdom predicts will increase in a warming climate.

Fire Regimes in the Western Mountains

In some western mountains, wildfires occur rarely, often separated by hundreds of years. In others, they can occur nearly every year. This variation is evident in both forests and other vegetation types. On average, the rarer fires are more *intense* and *severe*[8] than more frequent fires, and the total area burned is usually larger, although

---

[4] The first phrase in quotes is from one (or more) standard definitions in the literature. A classic treatise is an edited volume, S.T.A. Pickett and P.S. White. 1985. *The Ecology of Natural Disturbance and Patch Dynamics*, Academic Press. The second phrase is deliberately non-technical, and my addition.

[5] But later in this chapter and in Chapter 9 we will get to how these events interact with climate-change forcing.

[6] As in the spread of infectious disease. A sneeze (for example) provides the force, and dense enough populations the medium.

[7] For example, some will claim that the atmosphere is the medium for hurricanes, tornadoes, and other windstorms. Even further down that path, one could argue that the Earth's gravitational field is the medium for land movements.

[8] Fire "*scientists*" (who study the physical elements of fire) and fire *ecologists* (who study the biological and ecological elements) distinguish between *intensity*, basically the total heat produced by the fire, with its associated *metrics* (see Chapter 5), and *severity*, the damage caused, such as tree mortality, nutrient depletion, or erosion.

low-severity fires can continue for weeks and eventually cover large areas. Depending on three factors, weather, fuel, and topography,[9] the intensity of the same fire event can vary greatly over time and space.

A *fire regime* is the overall picture of wildfire occurrence, size, frequency, severity, and spatial configuration for some defined landscape over some defined period. In the parlance of research and management, frequency and severity are typically the defining features. For example, the western slopes of the Cascade Range, especially in the north, are said to have a low-frequency high-severity fire regime. Ponderosa pine forests in the Southwest are said to have the opposite: low severity but high frequency of fire. These two examples are the ends of a *gradient*, on which a fire regime's position is not always clear. More and more, we are coming to describe fire regimes in the West as *mixed-severity*, meaning that within the same area, or within the same fire, fire intensity and therefore its severity[10] are almost always variable.

I keep the cause of fire *ignition* separate from other aspects of the fire regime because some fire regimes are not affected much by how fires start, whereas some are affected critically. In the West, outside of urban areas and industrial structures,[11] wildfires are started by either humans or lightning. More area is burned by lightning-caused fires than by the human-caused, even though more individual fires are started by humans than by lightning. In a remote wilderness in which lightning is common, it may matter little to the eventual outcome whether a person or a lightning strike started a fire. In areas in which lightning-caused wildfires almost never happen, however, a human ignition is game-changing. For example, wildfires that occur in the Coast Ranges of southern California are started by humans, but the very high-intensity fires that burn through the chaparral and into communities are driven by the *Santa Ana* winds: hot, dry, and powerful easterly winds in late summer and fall. In a warming climate, this high-severity fire regime will depend on a combination of any regional-scale changes in the Santa Ana winds,[12] human carelessness, or arson, and demographic and regulatory changes.[13]

---

[9] These are the elements of the classic "Fire Triangle", which can be seen in many forms, both verbal and graphic, with an internet search on the term.

[10] In general, more intense fires are more severe, per unit area, for obvious reasons: flames, heat, and smoke kill organisms. This will not hold perfectly when organisms vary in their vulnerability. For example, large trees with thick bark can survive more intense fire than can seedlings, saplings, of mature trees with thin bark. See the discussion of adaptations below.

[11] Such as power plants. Obviously industrial accidents can start fires that can be intense and cause a lot of change in ecosystems as well as killing humans and damaging structures.

[12] And that is one of those regional-scale changes that we have a hard time modeling, so far. See Chapter 3.

[13] Discussion of these would take us far afield, but the so-called *wildland-urban interface* (WUI), land areas in which human structures, usually homes, and open-space vegetation are interspersed, sometimes in complicated spatial patterns, presents a huge conundrum around regulating development, landscaping within developments, and liabilities, all in the context of what seem to be unfailingly lucrative financial situations.

Gradients of Fire Frequency and Severity at Different Scales

Across the West, fire regimes are constrained by the patterns of climate and hydro-climate that I discussed in earlier chapters. Starting with westerly flow off the Pacific Ocean, each of the mostly north-south major mountain ranges has a wet and a dry side, caused by the rain shadow. Moving from west to east along the maritime-continental *gradient*, both the wet and dry sides become drier, and seasonal and diurnal differences in temperature increase. Almost all the "action" in wildfire in the West occurs when weather is warm and dry enough,[14] in contrast to other areas in the U.S. and around the world where other seasonal factors are important.[15]

At the broad scales of whole ranges, then, we see fire regimes that are associated with wetter climates on the west slopes of ranges, and those associated with drier climates on the east slopes. In the language of frequency and severity, which is which? In wetter climates, where vegetation is not *water-limited*, there is more biomass.[16] At the same time there is less opportunity for it to burn. So (once again on average) we see more frequent fires on drier east slopes than on wetter west slopes, but western-slope fires are more severe. In ranges with the greatest climatic difference between wet and dry "sides", such as the Cascade Range, fire regimes are the most different between west to east.

Implicit in what I am saying here is that on average, fires in water-limited systems are more frequent and less severe than fires in energy-limited systems. This is broadly true, but for understanding fire regimes in the West it turns out to be more directly useful to change our terminology a bit, while keeping our central idea of limiting factors. Let us distinguish *flammability-limited* fire regimes vs. *fuel-limited* fire regimes. The first type has no shortage of vegetation that could burn under the right conditions. Most forests are flammability-limited, with enough trees per unit area and sufficient understory vegetation under all but dense canopies. But these landscapes are not always flammable; much of the year they are too wet, too cold, or both. Examples of fuel-limited systems are sparsely vegetated savannas or woodlands, shrublands, and desert grasslands in which the vegetation is intermittent. They are almost always dry enough to burn, but often lack enough continuous fuel to carry a fire. In the next section we will see how these contrasting limiting factors help us to predict broad-scale patterns in fire regimes that will be useful for future projections in a warming climate.[17]

---

[14] For example, in the American Southwest, "warm and dry enough" begins in April or May, and "dry enough" ends with the arrival of the monsoon (see Chapter 4), even though temperatures remain high.

[15] For example, Florida, USA, where most wildfire happens in summer or winter, and the seasonally dry tropics, where the fire season is the dry season.

[16] Yes, it might be *energy-limited*, but for West-wide averages, water limitation is more pervasive, and more limiting, something that will only become more unbalanced as the climate warms (see Chapter 5). This logic therefore holds at the broad scales we are considering right here.

[17] For the enterprising reader, we wrote two papers about this. D. McKenzie and J.S. Littell. 2017. Climate change and the eco-hydrology of fire: will area burned increase in a warming western USA? *Ecological Applications* 27:26–36. J.S. Littell, D. McKenzie, *et al.* 2018. Climate change and future wildfire in the western USA: an ecological approach to non-stationarity. *Earth's Future* 6. https://doi.org/10.1029/2018EF000878.

At finer scales, topography becomes important in shaping fire regimes, both by affecting microclimates and by controlling the spread of individual fires. Fires will spread more easily and rapidly upslope than on level ground,[18] and more easily on level ground than downslope. There is a limit to this. For example, fires will not spread up or down cliffs, but in general they will spread up any slope that is gentle enough to have continuous vegetation to fuel the fire. Across a mountain landscape topography thus creates corridors for fire, but also barriers against it. The influence of both corridors and barriers on a particular fire depends on the current weather—temperature, humidity, wind direction and strength—and the vegetation in the fire's path (fuel). Clearly stronger winds, higher temperatures, and lower humidity make corridors (even) more passable and barriers less effective, but wind direction can swap these two features' effects. For a simple example, consider an area with mainly north-south ridgelines separating long valleys. A fire driven by westerly or easterly winds will encounter mostly barriers, whereas a fire driven by northerly or southerly winds will find corridors (Fig. 6.1).

You can understand why these fine-scale subtleties matter for practical consider-ations like firefighting or protecting habitat for sensitive species. For firefighting, understanding how topography and wind interact during a fire can save lives by helping firefighters anticipate in which direction and how fast a fire will spread. For protecting habitat, studying fine-scale topography helps to identify *fire refugia* (areas that "should" have burned when the rest of the landscape did but escaped it) that can be all-important for wildlife species that are sensitive to fire.[19] But does this fine-scale variation make things too complicated to estimate what will happen to fire regimes, at these scales, in a warming climate? It won't be easy; formidable com-plexities are associated with this particular scale. To understand them, we need more background, which will come later in this chapter and in Chapter 8. Before that, we need to bring in some other players: smoke and its consequences, other disturbances, vegetation (as not just fuel), and the interactions of all these things in a non-stationary climate.

---

[18] It is easy to see why this is so. Imagine a vertical flame on a slope. Every part of the flame except what is next to the ground is closer to the land surface on the uphill side than on the downhill side, making it easier for any flammable substance to ignite on that uphill side.

[19] For example, as most of us know, the Northern Spotted Owl was declared endangered as a result of massive clearcut logging in the Olympic Mountains and the west side of the Cascade Range, decimating its habitat. It turns out that population pressures (we will see more about this in Chapter 7) drove a considerable percentage of the survivors across the Cascade crest into marginal, but still livable, habitat on the eastern slope, which (typical of eastern slopes) has more frequent fire than the western side. It turned out that a large proportion of these dispersal sites had been fire refugia in historical times, but in an already warming climate they are now quite susceptible to wildfires severe enough to render them useless as habitat for the owl.

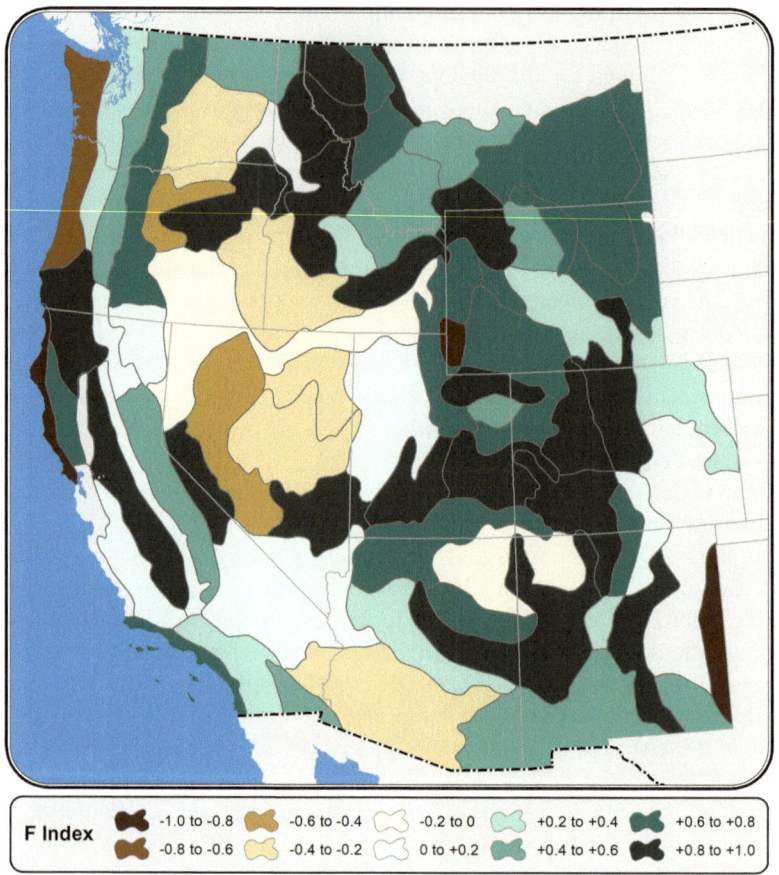

F Index		-1.0 to -0.8		-0.6 to -0.4		-0.2 to 0		+0.2 to +0.4		+0.6 to +0.8
		-0.8 to -0.6		-0.4 to -0.2		0 to +0.2		+0.4 to +0.6		+0.8 to +1.0

Fig. 6.1 Flammability versus Fuels as limiting factors. An "F" index, scaled to range between −1 and 1, measures how an ecological region is limited by flammability or fuels, on average. More flammability-limited = green (0–1), more fuel-limited = brown (−1,0). Across the West, regions will move from green toward brown in a warming climate. If a region changes from darker green to lighter green, it is likely to see more burned area, whereas if it changes from light green to light brown, or lighter brown to darker brown, it will see less burned area. We provide the calculations for the F-index in the papers listed in Note 17 for this chapter. Map by Robert Norheim

Smoke and Regional Haze

Smoke is one aspect of wildfire that seems to have no redeeming features, whether for humans or for ecosystems overall. Whereas fire is an integral part of the functioning of many ecosystems worldwide, smoke is a toxic by-product. For humans, smoke exposure via inhalation is acutely toxic and lethal at surprisingly small concentrations,[20] and it also has cumulative toxic (and sometimes lethal) effects. Cumulative effects are usually from small particulates,[21] which can enter the fine tissues of the lungs and remain there.

When fire is intense enough that smoke is lofted into the prevailing windstream, it can be transported hundreds of miles downwind. This is a *teleconnection* associated with fire regimes; an ecosystem process (fire) affects not only the area in which it occurs, but also distant landscapes. For example, smoke from wildfires or prescribed[22] fires in Washington and Oregon often reduce visibility greatly in Glacier National Park (Fig. 6.2), and fires in the Coast Ranges of California can reduce visibility in the Sierra Nevada (Fig. 6.3). More rarely, smoke from inland fires can reduce visibility and elevate air pollution in coastal cities, when anomalous regional wind patterns coincide with wildfires. During the 2018 summer, the San Francisco Bay area and Seattle, Washington, were engulfed by smoke from inland fires, such that on at least one of those days Seattle's air quality[23] was the worst globally of any reported urban measurements. Across the West, on average, the days with the worst air quality in the "Class 1" areas (national parks and other protected areas with air-quality monitoring instruments[24]) have a significant proportion of their air pollution, and reduction in visibility, caused by haze (smoke) from fires. With the *forcing* of wildfire by climate, which we turn to next, significant changes (reductions) in air quality across the West are likely.

---

[20] For example, it takes a surprisingly small percentage of carbon monoxide (CO), present in high concentrations in smoke, to replace enough oxygen to kill any organism. The "Immediately Dangerous to Life or Health Concentrations" (IDLH) for humans is said (by the CDC) to be 1200 ppm, but this is a general criterion and will vary among both people and circumstances (e.g. length of exposure, other ambient concentrations).

[21] Worst of these are the smallest, particulate matter less than 2.5 μm (a millionth of a meter) in diameter, called "$PM_{2.5}$". There is good medical evidence that $PM_{2.5}$ from fire is more toxic to humans than from other sources, even of the same size.

[22] I won't mention these much until Chapter 9, but prescribed fires are fires that are set intentionally by land managers to achieve some objective. Common ones are (1) to reduce fuels on a site so as to limit or prevent high-intensity wildfires, and (2) to eliminate or control some species of vegetation to promote others or increase the diversity of species overall.

[23] During the summers of 2017 and 2018, living in Seattle, I was asked by the media if climate change meant that all the summers from then on would be smoky. No, they won't all (and as I write (2019) Seattle is having its coolest and wettest July in years); weather will always be variable, and the smoke was a "perfect storm" of the coincidence of anomalous weather and wildfires inland. That said, if wildfire increases in British Columbia, Canada, in a warming climate, and regional circulations change, both of which are very possible, we could see more smoky days in a average summer, and more frequent summers with smoke invasions.

[24] This network of sites is called IMPROVE (Interagency Monitoring of PROtected Visual Environments). Their website has stunning visuals as well as data. http://vista.cira.colostate.edu/Improve/.

Fig. 6.2 Thick regional haze over Lake McDonald, Glacier National Park, transported from fires in Washington and Oregon. Photo courtesy of IMPROVE, US EPA. http://vista.cira.colostate.edu/ Improve/photos/ (see Note 24)

Fig. 6.3 Heavy smoke southwest from Yosemite National Park, with Half Dome in the distance. Photo by the author

Wildfire and Climate: How We Know What We Know

As with global warming and its forcing by CO_2, we have multiple lines of evidence for the forcing of wildfire by climate. We have two main types of paleological records, sediment charcoal and fire-scarred trees (see below), and many types of *observational* records (written down or entered in some way, as opposed to reconstructed). The paleological records sacrifice temporal and sometimes spatial resolution for record length; in the Western Mountains they extend back through the *Holocene* Period. Our observational records date back only to about 1900, with a range of climate variation that is much narrower than that of the entire Holocene.[25] Both types of records are invaluable. When the interactions of interest are complex and variable, one "sample" can be an anomaly, whereas many samples produce averages and variation around them.[26] More is better when it comes to the number of samples. For the paleological record, one sample can be 100 years, so many samples (e.g., 180,100-year periods over 18,000 years, the length of the Holocene) are needed. For the observational record, one sample can be how much area burns in 1 day, so 1 year may be enough time to gather many samples.

Sediment Charcoal

Paleontology takes huge advantage of how much of the Earth's surface has been deposited in layers, over time. Before we had precise dating methods that use radioactive isotopes, we could still examine sediments and know which layers came earlier—because they were farther down. The fossil record that tells us how life evolved over 3+ billion years is the best known example in biology. For wildfire, we take similar advantage of *varves* (annual layers of sediment or sedimentary rock) in lake sediments to date various fossils in the paleological record. Fossil charcoal is evidence of fire itself, and pollen and *macrofossils* (see Chapter 5) provide a record of plants and other organisms. By a judicious combination of radiocarbon dating and locating samples accurately in varved sediments, we can date sediment charcoal and associated other fossils to about decadal accuracy.

---

[25] Recall, though, that we are now "off the (Holocene) chart". Global average temperatures are higher than they have been in the Holocene.

[26] Technically, both fire history and climate simulations are what are known as *stochastic processes*. Each realization (instance that they happen) will be different because they don't start in exactly the same place. Fire history is "real", and climate simulations are "fake," but they both have this property. This is easier to see with a simulation; we start with *initial values* of parameters and these change when we run the simulation again. With fire history, in order to understand it properly, we proceed with our analysis AS IF it were only one instance, instead of its being the only instance that happened. If we have a long enough record, different chunks of time can be the "repeats," and we can infer, for example, the "average" influence of climate on wildfire over 100 years.

The relative abundance of charcoal at different depths informs inferences about the fire regime at the time, and the *fire history* over time. Many factors influence charcoal deposition, but if we look at the entire record, patterns of "peak" activity emerge against a "background". In the background, charcoal will still be present, but the peaks can be distinguished by various statistical means,[27] giving us an idea not only when fire occurred, but also how "much" (how intense, or how widespread, or both) fire was in the vicinity[28] of the lake. Putting together fire history, the history of vegetation change, from macrofossils and pollen, and reconstructed climate, we make "millennial-scale" inferences about the interactions of wildfire, vegetation, and climate. Not surprisingly,[29] hotspots for sediment-charcoal reconstructions in the Western Mountains are the Cascade Range and the northern Rocky Mountains including Yellowstone National Park.

Fire-Scarred Trees and Stand Reconstructions

Living trees, and their remains if they are more or less intact,[30] give us more detail about fire history than charcoal records, but the record extends into the past only to their maximum lifetimes. With rare exceptions,[31] this is hundreds versus thousands of years for sediment charcoal. We have already seen the remarkable value of tree rings, with their annual resolution. We use them in two ways to establish fire history, and to identify (usually) the exact year in which a fire occurred.

First, when fire reaches a tree but is not intense enough to kill it, it will often scar the tree near the ground. As the tree continues to grow, it will heal the wound and gradually close around it, leaving the scar tissue (literally) hidden in place (Fig. 6.4). In *low-severity* fire regimes, trees may have multiple fire scars, and each can be

---

[27] To do these statistics right is not for the timid, and it also requires care in avoiding an over-reach of inference. For the interested, a moderately deep dive into the methods is P.E. Higuera, *et al.* 2010. Peak detection in sediment-charcoal records: impacts of alternative data analysis methods on fire-history interpretations. *International Journal of Wildland Fire* 19:996–1014.

[28] This vicinity can be defined only imprecisely, because charcoal fossils can be deposited some distance from the combustion that produced them. Wind, topography, and fire intensity all contribute to the variability in distance traveled. This is OK. In rare cases we might like to reconstruct the perimeters of specific fires, but normally these data are used for broader inferences.

[29] Can you guess why? Which level of fire severity—low, high, or mixed—would be easiest to detect in the charcoal record? Hint: the northern boreal forests are another region with numerous sediment-charcoal records.

[30] Stumps, logs, or *snags* (dead trees still standing) whose tissues are preserved enough for tree rings to be evident.

[31] For example, giant sequoias 3000+ years old in the Sierra Nevada, and bristlecone pines 4000+ years old in the White Mountains, a Basin Range (see Chapter 2).

Fig. 6.4 Fire-scarred giant sequoia in Sequoia National Park. Photo by the author

dated by its position in the tree's growth rings.[32] A study of fire history will typically look at an area large enough to sample enough trees so as not to miss any fires within that area in the lifetime of the trees. This history can extend back several hundred years, at annual resolution.[33]

Second, in high-severity fire regimes, we can still take advantage of the annual resolution of tree rings. In forests, after a *stand-replacing* fire (killing most or all trees), young trees will re-establish over a period from years to decades. If most or all fires are stand-replacing, a spatial pattern of *age classes* (groups of trees, "patches", on the landscape that are the same age (sometimes exactly) or similar ages. By selecting a large enough area, and measuring the ages of enough trees in

[32] I am making this sound straightforward, for our larger purpose, but this process can be quite delicate. Rings in some trees can be missing, but not in others nearby, so that with many trees, if one is not careful the number of years with a fire scar can be over-counted. Recall again that "absence of evidence is not evidence of absence." Trees that experience a fire will not always be scarred. In particular, the first scar is the hardest to make.

[33] And some researchers have managed to place fire scars precisely enough within the annual growth cycle so as to estimate the season in which a tree was scarred. This is impressive, but we won't need the details for our discussion here.

each patch to have a good estimate[34] of the date the patch originated after fire, we have a map of the most recent fires that affected each patch. This is not as fine a record as that from fire-scarred trees, but with a large enough sample[35] it still can be surprisingly informative about the association between climate variability and wildfire. In the end, these *stand reconstructions* give us a resolution "in between" those of sediment charcoal and fire-scarred trees.

Not all fire regimes are purely low-severity, with all trees surviving, or high-severity, with all trees killed. In recent years we have seen creative combinations of these two tree-aging methods[36] to establish metrics for *mixed-severity* fire regimes, which are the norm in many forests of the West.

Native American Burning: Confounding the Inference?

Beginning ~11,000 years BP, Native American cultures around the West set fires regularly, often to maintain wildlife habitat and maintain a mosaic of forests and open grasslands to facilitate hunting and travel and encourage growth of plants for food and other uses. This history of controlled fires overlaps, in both time and space, the complete record of fire-scarred trees—both essentially ended by 1900 and occurred on the same landscapes. But the existence of controlled burns does not invalidate inferences about the effects of climate on wildfire. In two ways it strengthens them. First, any fire set for a specific purpose becomes "noise" (in the sense of signal-to-noise) in the signal from fire climatology, being not *correlated* with any particular period of climate either conducive or unfavorable to fire.[37] Second, there is good reason to believe that Native Americans sometimes used controlled fires in one way that fire managers do today, to reduce the risk of large severe fires by

---

[34] This good estimate can be tricky to verify. Because trees establish at different times after a stand-replacing fire, the end result will be a *distribution* of ages for each patch. A reasonable choice for the presumed year of the fire is the assumed establishment date of the oldest tree ("fire was no later than this year"), but there are arguments pro and con other choices.

[35] For a toy example, consider a fairly large landscape, probably sampled coarsely by *remote sensing* (images from aircraft or even satellites) to define the patches, and then by hand to get the tree ring counts. Suppose, for simplicity, that all patches were the same size, and that the following patch ages were estimated: 500, 475, 450, 425, 400, 250, 100, 75, 50 years since fire. It would be reasonable to assume that there was more fire in the early and late parts of the record, and less in the middle. Knowing the climatology (e.g., the *Little Ice Age* was in the middle of the period), we have the basis for inferences about the effects of climate on wildfire in this region.

[36] Some good overviews of this work are J.E. Halofsky, *et al.* 2011. Mixed-severity fire regimes: lessons and hypotheses from the Klamath-Siskiyou Ecoregion. Free online from *Ecosphere* 2(4):art40. doi: 10.1890/ES10-00184.1. P.Z. Fulé, *et al.* 2003. Mixed-severity fire regime in a high-elevation forest of Grand Canyon, Arizona, USA. *Landscape Ecology* 18:465–486.

[37] At least not in any clear way that can be incorporated in the analysis. For example, would there have been more controlled burns, or fewer, in a hot dry summer conducive to large lightning-caused fires? We don't know.

reducing fuel. If so, these human-set fires would remove some statistically "important" data points from analyses of climate and wildfire, by taking out the big fires that (presumably) occurred in years of fire-conducive climate.

The Observational Record

We are fortunate that human record-keeping about wildfire began at about the same time that the paleological record ends. Around 1900, fire *exclusion* (any method for keeping fires out of ecosystems) was becoming effective enough that the fire-scar record all but ends.[38] Not long after that, fire occurrences, locations, and sizes began to be recorded systematically. We have about a century-long observational record of wildfire, as we do with climate. Both have gradually improved over the years. In the 1980s, the Landsat satellites started delivering images at 30-m resolution that could be interpreted to estimate fire severity. Most of the larger wildfires in the West, those more than 1000 acres, are part of an online database called "Monitoring Trends in Burn Severity" (MTBS: https://www.mtbs.gov/), which is heavily used by fire researchers. Besides these West-wide observations, all fires wholly or partly on *public lands* (those managed by the US Departments of either Interior or Agriculture) are recorded in agency databases.

We have a lot of information about wildfires in the last 100 years, and more details for more recent fires. Because the same is true of climate and weather, there are many different ways to infer how climate drives wildfire from these data. Three important ones are useful at different scales.

At broad spatial and temporal scales, we develop statistical models of how annual or seasonal fire area, in a region of interest, is influenced by different climatic variables such as temperature, precipitation, the water balance deficit that we met in Chapter 4, and others.[39] These models are *static*, in that they are snapshots in time.

At fine spatial and temporal scales, we estimate where a fire will spread, how fast, and how intense it will be, using models of *fire behavior* (what fire does) that invoke some of the same climatic variables, albeit at fine scales. These models are *dynamic*, in that they replicate or simulate a process over time.

---

[38] For example, probably the most detailed and comprehensive fire-scar record we have in the West is from eastern Washington State. In every one of seven watersheds sampled, trees have "stopped" recording fire by 1900.

[39] But as in other disciplines, some of these variables are fake. A classic, and a nice example of "the tail wagging the dog," is the *fire season*, and its length. Papers have been published in prestigious journals, and cited profusely, that make claims like "Wildfires are increasing around the West because the fire season is getting longer". But this is circular logic. The fire season is defined or bounded by the fires that occur, it does not cause them, any more than being in the rainy season causes rain.

At medium scales, we develop *spatially explicit* (it's important where things are in relation to other things) models of associations between fire severity and topographic position, then contrasting this with the effect of climate on the patterns of severity and deciding which is dominant.

These and other analyses have limitations in the context of projecting fire climatology in a warming climate. We will see some of those in the section "Landscape Disturbance Models"below.

Bark Beetles and Other "Winners", and Interactions

Metabolic rates in living organisms share one property with combustion in wildfires: they increase with heat. Organisms are responding to a globally warming climate, both negatively and positively.[40] Those with positive responses are taking advantage of rising temperatures by colonizing new habitat,[41] increasing populations in current habitat, or both. For the southern United States, for example, there is concern that tropical diseases like malaria and other mosquito-borne illnesses could gain a foothold in a warming climate because mosquito populations will increase. For the Western Mountains, there are similar concerns,[42] but our focus is ecological disturbance, and I will stay with an example whose ecological effects are more wide-ranging and evident, even now.

Native insects and *pathogens* (named so because they kill things that we care about)[43] have been a part of the Western Mountains throughout the Holocene. Of the two, insects are more visible and more "disturbance-like" because their principal damage is through *outbreaks* rather than *baseline mortality* (i.e., continuous). An outbreak happens, just as with human diseases, when some *limiting factor* is relaxed or overcome.

---

[40] Right here we are focused on the "winners". In Chapter 7 there will be more about some "losers".

[41] Of course there are two ways to do this. What has little to do with warming climate directly is the success of invasive species, both plants and animals, such as cheatgrass across much of the West and buffelgrass in the Southwest, which were given a running start by being transported here when they would have not dispersed so far on their own. The second way, dispersing from adjacent habitat by taking advantage of a new more favorable climate, is the one I will discuss.

[42] For example, the mosquito whose bite carries dengue fever, *Aedes aegypti*, could have about six times as long a breeding season in middle-to-high elevations in the West by the end of the century, in a "business as usual" scenario for climate change. For details see S.J. Ryan *et al.* 2019. Global expansion and redistribution of Aedes-borne virus transmission risk with climate change. *PLOS Neglected Tropical Diseases* (a free online journal) 13(3), https://doi.org/10.1371/journal.pntd.0007213.

[43] And we won't deal with these, for various reasons, including their sheer number and variety and a lack of confidence, for most, that we know what will happen with a warming climate. For some good information on pathogens, though, see https://www.fs.fed.us/research/invasive-species/plant-pathogens/, which covers both invasive and native pathogens.

Climate (e.g., cold temperatures) can be a limiting factor, as can some predator, such as a bird species. In the evolutionary game, trees have evolved defense mechanisms that can be fairly general or in response to one species that is encountered most. These defense mechanisms are another limiting factor; usually when they fail it is because they are overwhelmed by sheer numbers of attacking insects.

Outbreaks of insects that kill trees are a major disturbance in the Western Mountains, in some years affecting more area than wildfires. Two major types are *defoliators*, which consume foliage, sometimes enough to kill a tree, and *cambium-feeders*, which bore into the living part of a tree bole, forming *galleries* for depositing fertilized eggs. When young insects emerge, they feed on the live tissue, usually killing the tree. Both types of insects "specialize" on one or a few species, so that when an outbreak occurs, there can be 100% mortality in single-species stands of trees, but patchy mortality in mixed-species stands.

The insect species whose response to climate change that we understand best are the bark beetles, cambium-feeders that attack pines. The West has the mountain pine beetle, the western pine beetle, and the pinyon Ips, or "engraver", beetle. These species have favorite *hosts* (the tree they attack the most), but are capable of switching, either opportunistically or under stress. The mountain pine beetle (MPB) will be the most illustrative for us, because we have seen over recent decades how its *life history* (literally, how it spends its life) is linked to temperatures and the seasonal cycle.

The preferred host of the MPB has been lodgepole pine as long as their "insect-host" interactions have been observed.[44] In the Western Mountains, and in interior British Columbia north of them, single-species forests of lodgepole pine cover vast areas. In recent decades, MPB outbreaks have hit much of these forests, leaving almost complete mortality of mature trees, from isolated stands to entire vistas (Fig. 6.5). The extent and severity of these outbreaks are driven partly by climate, and partly by *host vulnerability* (something about the host that makes it particularly susceptible). That "something" is simply timing. The forests that were attacked were almost uniformly of mature adult trees. These are the most attractive to the beetle, whether or not the climate is favorable. So the severity of an outbreak is attributible to both climate and the age of the trees.[45]

The climate factor comes from the degree of synchrony between the MPB's reproductive cycle and the seasonal cycle. As I stressed earlier, metabolic rates of insects increase in a warming climate. It turns out that the reproductive cycle is controlled strongly by metabolic rates; the warmer the average temperature, the faster the insect matures through a series of larval stages and emerges from its host tree (having killed it, with help from gallery "mates"). Having evolved with the seasons in a temperate climate, however, constraints are built into these changing

---

[44] We don't have any particular reason to believe that has changed over time, but read on.

[45] How much of each? The alert reader may see the opportunity for a detection and attribution lesson or analysis here, which could provide valuable data for a land manager. Our focus being on climate, we have something different to emphasize, but if you want to drill down, think *limiting factors*.

a

b

Fig. 6.5 Lodgepole pines killed by the mountain pine beetle. (**a**) The "red stage", about two years after the outbreak. Photo by Alina Cansler. (**b**) The "gray stage", 3+ years after the outbreak. Photo by Jeffrey Hicke

Fig. 6.6 Whitebark pines, at treeline, killed by mountain pine beetle. Photo by Jeffrey Hicke

rates. The "normal" state for the MPB, and other insects, is to be more or less synchronized with the seasons, such that it emerges at the most favorable time of the year for reproduction. This is called *adaptive seasonality*, and outbreaks depend on the reproductive cycle being "centered" on some multiple[46] of a year, but they are much worse if that multiple is one than if it is two or more. This is what has happened to the MPB over much of the range of lodgepole pine.

There is another "win" for the MPB besides its halved generational turnover. Its "effective" range (for reaching numbers lethal to hosts) has expanded into forests of whitebark pine, a charismatic and key species in high-elevation forests (Fig. 6.6). This is simply global warming, exacerbated by the elevation-dependent warming that we saw in Chapter 3. As we turn next to disturbance interactions, we will see how whitebark pine is in double jeopardy.

Disturbance Interactions

Most ecosystems in the Western Mountains experience more than one type of disturbance. These disturbances often interact, and ecological interactions can be *synergistic*, *antagonistic*, or some of each. How much of each will depend on the timing, extent, and severity of the two (or more) disturbances. For example,

---

[46] This is a fair amount of technical detail already, and there is more in the full story. There are relatively stable "states" of the MPB's cycle; it has been biennial in the past, but temperatures have warmed enough in large areas of its range that it has switched to annual, meaning that a new cohort of beetles is produced twice as fast as before. These are the sheer numbers that overwhelm the host trees. For the technical details unabridged, see J.A. Powell and J.A. Logan. 2005. Insect seasonality: circle map analysis of temperature-driven life cycles. *Theoretical Population Biology* 67:161–179.

consider a bark-beetle outbreak before a wildfire. Soon after an outbreak that is severe enough to kill most trees in a stand, the needles[47] will desiccate, making the tree canopy more flammable, and a wildfire will be more severe over the area of the outbreak than it would have been with no beetle attack. But within a few years[48] the needles will have fallen off branches, leaving much less fuel in the canopy than there had been, and reducing fire severity.

If one disturbance among some that interact changes in a warming climate, then clearly so will the interactions. In the Western Mountains, the interactions most likely to change[49] in the face of future warming are indeed those between wildfire and insect outbreaks. I shall continue to focus on bark beetles, because their example has the clearest link to warming temperatures, but there are certainly other insects that can reach outbreak numbers in the Western Mountains, and whose numbers are strongly affected by climate.[50]

What will we see in the Western Mountains where we expect wildfire and bark-beetle outbreaks to continue in a warming climate? I have already pushed back, and will do so again (e.g., in Chapter 8), on the idea that warming climate will bring more wildfire everywhere across the West. One place on which there is agreement,[51] however, is medium-dry to dry forests, which are expected to be warmer but not wetter in the future. These also happen to be a domain of the MPB, which is in the process of changing, as we saw, from a *semi-voltine* to a *uni-voltine* species (a brood of offspring every 2 years to 1 every year). Our naive belief may be that "more fire and more beetles = more *synergistic* interaction = more area affected, more severely". This will certainly happen in some places, but remember our example of timing. If enough time (just a few years) passes between an outbreak and a fire, fire may be less able to spread and therefore affect less area, less severely. Conversely, if a fire happens first, the population of vulnerable hosts for the beetle will be reduced,[52] and an outbreak will be less sustainable.

Beyond the complexity of this interaction in a single case (one fire, one outbreak), *negative feedbacks* are associated with increasing frequency and extent of the interaction. Stands of forest take time to be vulnerable, to both fire and insect outbreaks. In the case of lodgepole pine and the MPB, for example, this can be as

---

[47] Remember that bark beetles attack pines. With broadleaf trees, this example is less clear.

[48] For much more on this topic, see J.A. Hicke *et al.* 2012. Effects of bark beetle-caused tree mortality on wildfire. *Forest Ecology and Management* 271:81–90.

[49] With respect to the classic elements of disturbance: frequency, severity, and (particularly) extent.

[50] For example, the western spruce budworm is a *defoliator* that has reached outbreak numbers in the Southwest nine times in the last 300 years. Unlike bark beetles, however, its strongest climatic driver seems to be precipitation; in the Southwest wetter springs are associated with increased activity. For the classic study, with our invaluable *tree rings*, see T.W. Swetnam and A.M. Lynch. 1993. Multicentury regional-scale patterns of western spruce budworm outbreaks. *Ecological Monographs* 63:399–424.

[51] Full disclosure: here is one forest type for which I mostly agree with the "universalists" who believe that wildfire will increase everywhere.

[52] This is not universal logic for bark beetles; it applies only to *live-cambium* feeders. Other species colonize wood that is already dead, and so the logic changes.

much as 50 years. So we are building the pieces of a complex puzzle about the future of disturbances and their interactions in the Western Mountains.

Some interactions can be lethal, of course, even if not all the disturbances involved are increasing, whether from warming temperatures or something else. As I mentioned, whitebark pine, an iconic high-elevation species that happens to be adapted to fire and is also an important resource for grizzly bears and other wildlife, is now being affected by MPB outbreaks. If that weren't enough, an *introduced* (i.e., by humans, not native) fungal pathogen, white pine blister rust, is killing whitebark pines across the West, in Crater Lake, Glacier, and North Cascades National Parks. Only a very small proportion of trees is even somewhat resistant to the fungus. Technically these two disturbances are *additive*,[53] not truly synergistic, but their combination is a major threat to the species' persistence.

Feedbacks to Climate

The most obvious and direct feedback from these ecological disturbances is a positive one. Combustion of biomass releases CO_2 into the atmosphere, along with other gases and particulates, thereby warming the Earth by the greenhouse effect. Across the West, warmer temperatures are likely to increase wildfire extent, on average, though not everywhere, as we said earlier. It is the warming, not the CO_2 itself, that is the *positive* (accelerating the underlying phenomenon) feedback.

Wildfires also produce aerosols, both particles and liquids, which we saw in Chapter 3 are one of the large uncertainties in GCMs and in predicting future warming in general. Depending on chemical composition, concentrations, optical properties, location (where in the atmosphere), and complex interactions with clouds, aerosols can either increase or decrease the radiative forcing that is responsible for warming. Some particulate types clearly produce a positive feedback to warming, however, when they return to Earth. When black carbon and mineral dust from wildfires are deposited on snow or ice, they decrease the albedo of the land surface greatly, thereby increasing the absorbed radiation.[54] This accelerates both overall global warming and *elevation-dependent* warming, the latter because of the greater proportion of snow and ice cover at higher elevations.

Feedbacks to climate from insect outbreaks are harder to identify. Because there is no immediate pulse of CO_2 like that with combustion, the feedbacks are a complicated mix of positive and negative: CO_2 released by decomposition of dead biomass (+), albedo changes from loss of tree cover (−), and accelerated uptake of CO_2 from rapid regrowth (−).

---

[53] In other words, changes in one don't affect changes in the other. For example, there is no particular evidence that it takes both of them to kill a tree; either one will suffice.

[54] A clear example of this, though somewhat technical, is Kaspari *et al*. 2015. Accelerated glacier melt on Snow Dome, Mount Olympus, Washington, USA, due to deposition of black carbon and mineral dust from wildfire. *Journal of Geophysical Research: Atmospheres* 120:2793–2807.

Avalanches and Landslides

Mass movements of snow, ice, and rock (avalanches), or the entire land surface (landslides), are limited in size physically by the topography, such that they typically cover less area than the largest wildfires or insect outbreaks (Fig. 6.7). Still, in some areas of the Western Mountains, they can equal or surpass these other disturbances in annual area affected, by their frequency. In the American West, they are intermediate in the amount of human habitation and activity they affect between wildfires (the most) and insects (practically none). In other areas around the world, e.g., the Alps, avalanches are a greater threat and have more effect on humans than wildfires.

Hydroclimate and topography are the keys,[55] for our purposes, to understanding mass movements and how they might change in a warming climate. Because topography does not change,[56] we will focus on hydroclimate. Both avalanches and landslides depend on instability in the material involved, which is a function of its cohesiveness and the gravitational energy available. For the land surface itself, on a slope of a given steepness, the greatest destabilizer is water. The worst landslides, in terms of fatalities, are rapid-moving events triggered by extreme rainfall. We can then infer, for example, that if the frequency and intensity of *atmospheric rivers* increase across the West, on average the likelihood of a particular landscape having a landslide will increase.[57] Predicting specifically where landslides might increase is a more difficult challenge, analogous to the increased complexity of predicting local vs. global climate change.

Our basic take-home message for landslides is that more rain (precipitation), especially extreme events, means more instability and more landslides. For rock avalanches where there is little snow or ice, the message is the same. When snow and ice are involved, more precipitation may mean more or less rain or snow, depending on the season and elevation of any location. Atmospheric rivers usually involve warmer than average air temperatures, so extreme rains at lower elevations turn into very wet snow, mixed rain and snow, or just rain at high elevations. Similar to landslides, avalanches are more likely after heavy snowfall, but there are complicating factors, particularly the history of precipitation over days, weeks, and even

---

[55] Earthquakes create some of the biggest mass movements, of course. Even if we someday find a way to predict earthquakes, projecting their changes in a warming climate is likely to remain beyond our reach.

[56] OK yes, mass movements change the specifics of local topography, but not its general characteristics, e.g., steep and rugged vs. gentle or flat.

[57] On average, and all other things being equal, which they never are exactly. For example, if a slide-prone slope gains tree cover over time from growth accelerated by warming, its inherent stability will increase, offsetting its increased annual likelihood of failing because of more frequent severe rainfall. For a technical exposition of the expectations under warming climate, see S.L. Gariano and F. Guzzetti. 2016. Landslides in a changing climate. *Earth-Science Reviews* 162:227–252.

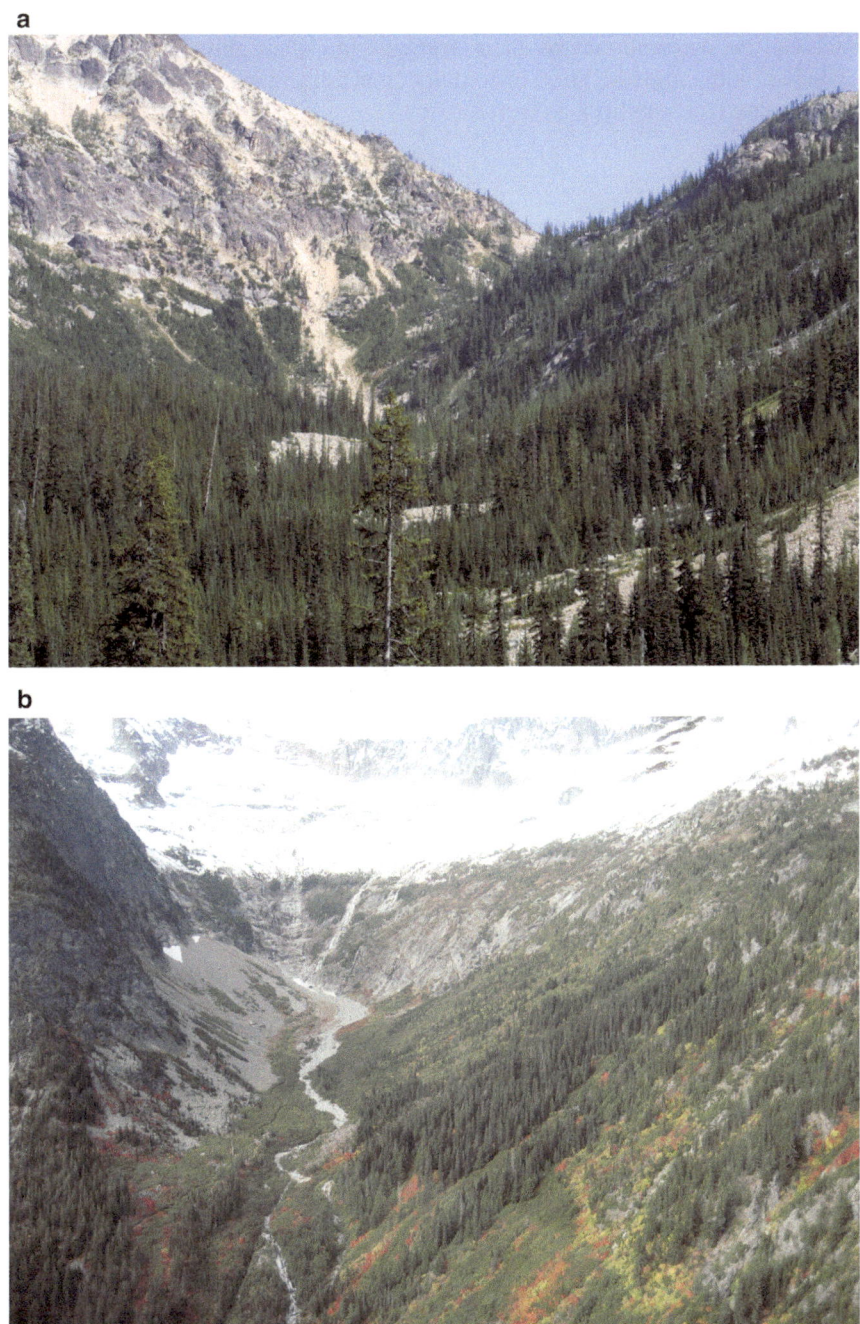

Fig. 6.7 Avalanche tracks in the northern Cascade Range, wherein they disturb more total area than wildfires or insects. (**a**) Photo by David L. Peterson, (**b**) photo by Jon Riedel

the season.[58] Consequently, the *variation* in temperature and precipitation over a winter can be as important as their averages in determining the likelihood of avalanches, with rapid alteration of cold dry snowfall and warm wet snowfall being the worst case (for stability).

Windstorms

The Western Mountains are not in the direct path of hurricanes, and most tornadoes are very local in effect, and much more common out on the Plains and further east. Damaging windstorms across the Western Mountains are more likely to be of two types: winter winds caused by extra-tropical cyclones in the eastern Pacific Ocean, affecting the Pacific Northwest, and anomalously warm (even hot) "Chinook" winds, which come in several varieties. The eponymous Chinooks are warm westerly winds that blow off of the Colorado Front Range down onto the Plains, and arise from the large temperature difference in winter from north to south across North America. The *Santa Ana* winds are hot easterly winds from the Southwestern desert that strike the Coast Ranges in Southern California, bringing hot temperatures in the autumn, when vegetation is the driest. In the Wasatch Range of Utah, the Chinook winds are called *foehn*, a name taken from similar winds in the Alps, and are easterly, blowing down the west-side canyons onto the Great Basin. All Chinook winds are themselves warm, and their rapid altitudinal descent heats the air compressively, raising temperatures further. The most severe Chinooks can reach 100 mph!

In the Pacific Northwest, in the Cascade Range and Olympic Mountains, the winter windstorms can level mature forests; large old trees are the most vulnerable, having the widest "sails". This *windthrow*, or *blowdown*, leaves tons of large logs on the forest floor, opening the canopy to younger trees that had been suppressed by the low light under the mature canopy. Over time, after some decay, these logs become *nurse logs*, the best substrate for tree seedlings and plants to establish and grow on a forest floor that is otherwise thick with vegetation.

Chinook winds can also blow trees down, but their biggest contribution to disturbance regimes is in fanning the flames of wildfires. As I mentioned, human-caused fires combined with the Santa Ana winds are a lethal mixture in Southern California. The Chinook winds on the Colorado Front Range can also drive wildfire through the low-elevation forests on the east side into nearby communities.

---

[58] For example, backcountry snowsports enthusiasts are familiar with the complexity of the layering of sequential deposits of snow in avalanche-prone terrain. Depending on their depth, moisture content, and the pressure from layers above them, "weak" layers (i.e., prone to slide) can persist under what may seem to be stable snow.

Disturbance and Succession in a Changing Climate

For ecosystems in the Western Mountains, ecological disturbance is a big player in the evolutionary game. Over millennia, the adaptations of plants and animals to fire and to insect and pathogen attacks affect their success at maintaining viable populations where they are, dispersal, and establishment of new communities, in the face of competition from other organisms. For example, pines and other tree species have adapted to repeated bark-beetle attacks by producing pitch, which (as it is named) literally "pitches" invading bugs out of the bark when they attempt to bore in to build galleries. For every defense there is a cost, however. Resources (energy, water, and nutrients) used to create pitch are not available for growth or building root systems. If a tree's growth is diverted for defense, it is more easily shaded out by others.[59]

In most mountain ecosystems, the tradeoffs associated with the presence of disturbances are often more complex than this example. Continuing our emphasis on forests and trees,[60] we have already observed how a tree species' place in the domains of the basic limiting factors, energy and water, determines where across the West it will thrive, survive marginally, or be absent. Disturbances, fire and insect outbreaks in particular, are certainly linked to the *hydroclimate*, and therefore the interplay of limiting factors, but they are nearly ubiquitous in the Western Mountains. For example, the only Western-Mountain ecosystems that have no fire are those with no vegetation at all; even the temperate rainforests of the Pacific Northwest burn occasionally. How trees, and plants in general, adapt to fire illustrates how disturbance complicates a species' response to the hydroclimate.

Adaptations to Fire: "Strategies" for Staying in the Evolutionary Game

Above a certain fire intensity, wood and leaves burn. No structure that a tree or other plant could evolve will protect it against being killed by fires that are severe enough. In contrast, with enough protection trees in particular can survive low- and even moderate-severity fire. Knowing this helps us understand how species with different

---

[59] In this example, a different species that is not vulnerable to the insect and therefore has not "spent" resources on the same defenses.

[60] These are the "actors" at center stage in the climate-change story. Other plants and microorganisms are certainly important in forests and other vegetation types, and we will look at the animal kingdom in Chapter 7. The erudite reader must forgive me for keeping the "casting" at a manageable scale.

adaptations[61] to fire are associated with different severities in the fire regime. So species associated with low-severity fire regimes have evolved more direct strategies, whereas species in higher-severity fire regimes have more evasive strategies.[62]

The main direct strategy is thick bark, which serves simply as a shield to protect the inner structures from lethal heat. Conifer species in the Western Mountains that have evolved thick bark are ponderosa pine, which is found all over the West, and two with more restricted ranges, western larch in the inland Northwest and Jeffrey pine in the Sierra Nevada and the central Pacific Coast Ranges. Other species, such as the wide-ranging Douglas-fir, have bark that is thick enough to provide moderate protection. Trees vary within and across species in the rates at which bark expands, such that it is large mature trees that have the thickest bark and can resist fire the best. These are also the trees that are tall enough that enough of their foliage escapes being killed directly by fire.

We see three evasive strategies in tree species of the Western Mountains. The most widely used is simply not being in the wrong place at the wrong time. These are species associated with later stages of *succession* that have no adaptations to fire *per se* but will slowly recolonize burned sites, and then persist until the next fire. As you can imagine, these species are associated with infrequent higher-severity fires. Classic examples in the West are the high-elevation species mountain hemlock and subalpine fir, and western hemlock, a *late-successional* species that is widespread in the Pacific Northwest. A second strategy is to ensure that the *propagules* (i.e., the next generation) are not burned, by sequestering long-lived seeds in the soil or via *serotinous* cones. These cones will open, distributing propagules, only under extreme heat from high-intensity fire. The classic example of *cone serotiny* in the West is the widely distributed lodgepole pine, an *early-successional* species that will repopulate sites on which most or all trees were killed, such as in the notorious Yellowstone fires[63] of 1988. A third strategy is to preserve the means of reproduction, if not propagules themselves. Some oaks and other deciduous species can re-sprout from roots after everything aboveground has been killed by fire (or something else).

---

[61] A key point here, which I come to below, is that adaptation to an environmental factor requires *evolutionary time*. In this example, the fire regime must be stable enough temporally in its severity for evolution to take its course. If fire regimes change quickly in a warming climate, many "bets" are off.

[62] For those who like classifications with evocative names, a researcher named J.S. Rowe wrote a chapter in 1981 in a now out-of-print book that named five classes: invaders, evaders, avoiders, resisters, and endurers. I would call all of these except resisters *evasive* strategies. For more details, track down R.W. Wein and D.A. Maclean. 1981. *The Role of Fire in Northern Circumpolar Ecosystems*. Or see James K Agee's classic *Fire Ecology of Pacific Northwest Forests*. 1993. Island Press. pp. 135–136, although the entire book is worth a careful read.

[63] We can see a subtlety in the evolutionary game by looking at how severe are fire regimes where lodgepole pine is dominant or common, and asking what percentage of the cones at each site are truly serotinous. Not surprisingly, the percentage of serotiny increases with fire severity, although nowhere does it reach 100%.

Evolutionary time for long-lived species moves more slowly than environmental change. So the more rapid the environmental change (e.g., climate warming at an unprecedented rate), the more "out-of-sync" is a species *fitness* (in the sense of Darwin) with its current environment. If fire regimes (and other disturbances) change rapidly, we can expect winners and losers. For example, if higher fire severity is the norm, what I called direct adaptive strategies, which evolved with fires that could be withstood, e.g., with thick enough bark, will start to fail. Analogously, if insect outbreaks of some species increase, their host trees will be compromised with respect to others that are not targets of the insects. In a worst case, with forests in which such a compromised (by fire or insects) species accounts for most of the trees, the continued existence of the forest itself is at risk.[64]

Evasive strategies—the ones in place now—are not necessarily winners either. Consider our first strategy of essentially waiting out fires and recolonizing sites "later" in succession by species that have no true adaptations to fire. This does not work if there is no "later". If fires were to become more frequent[65] on sites on which late-successional species eventually dominate, that eventuality could cease, with fire always interrupting and "resetting" the late stages of succession. On the other hand, at first glance cone serotiny would seem to be a clear winner, because fire effects the regeneration of serotinous species directly. But this can happen only if there are cones, which are only on mature trees. If a site reburns before trees mature, or there are not enough mature trees that survived the previous fire, there is no seed source.

The serotinous lodgepole pine and the high-elevation whitebark pine may have particularly interesting and complex futures. Lodgepole pine's adaptation to high-severity fire would seem to give it a competitive advantage over late-successional species in a warming climate, but because it is the preferred host of the MPB, the opposite happens in a forest where other species are unaffected. Where both disturbances are part of the ecosystem, we can be sure that the timing of disturbance interactions will be important, and we can expect multiple *successional pathways* (see Chapter 5). We already noted the double jeopardy of whitebark pine to the pathogen *blister rust* and the invasion of higher elevations by the MPB. Whitebark pine colonizes sites after fire, however, so more fire is to its advantage, both directly and if it regenerates more sparsely after fire,[66] thereby muting the effects of the other disturbances, which thrive on denser host populations.

---

[64] See just below where we revisit "Forests on the brink."

[65] Forecasting changes in fire frequency is not straightforward, despite claims in the media and elsewhere that a warming climate will cause more frequent fires. Consider the *negative feedback* to fire frequency from more area burned and more severe fire (both also being predicted for the future), if fire renders a site much less flammable, even if for only a few years. More on this later in Chapter 6, in the section on detection and attribution.

[66] Fire does kill whitebark pines, but their seeds are often cached by birds (see Chapter 7) some distance from their source, thereby varying the locations of seedlings.

Regeneration: Enabled but Vulnerable

Seedlings and young trees have a narrower range of environments that they can tolerate than mature trees. If it is too cold or hot, or too dry or wet, seedlings in particular will die when mature trees may be just slightly stressed. In undisturbed forests, we can expect a lag effect of a warming climate. For example, water-limited trees can survive decades or even centuries after the climate has warmed to one that favors species that are adapted to a hotter drier environment, and long after their seedlings would not be expected to survive, either from being killed outright or being outcompeted by other species.

Disturbance is a game-changer. If severe enough, it can interrupt the life spans of those mature trees that otherwise would have persisted for many years. As the forest landscape is opened to colonizing species, the successful ones will be those that are adapted better to the current environment. The new vegetation will thereby be more "in sync" with the current climate. This process works only when there are new species in the "pipeline"; a seed source must be nearby and the species has to tolerate the new climate in its most vulnerable (seedling) life stage. For example, in a forest populated entirely by *late-successional* species (shade-tolerant trees that grew initially under a canopy) with *evasive* strategies to adapt to disturbance, these species may be the only remaining seed source nearby, and unable to tolerate the new climate.[67] If no species are in the pipeline, a forest may indeed be "on the brink" (see below).

Fire-Induced Changes in Treeline, No Simple Fate

The subalpine "parklands" at our upper treelines in the Western Mountains are some of our iconic landscapes. Most of the mountain ranges I described in Chapter 2 have these areas, which are disproportionately[68] large parts of national parks. They are worth a short detour here as an illustrative example of the complexities of the interactions among disturbance, succession, and a changing climate.

---

[67] There are subtleties here that I am not bringing in, to keep the narrative clear. Late-successional species' adaptations to the climate that was in place when they were seedlings would not necessarily extend to regenerating successfully in the open; even then they need the moderating environment under a canopy. If heat or drought were the "worst" parts of that open environment, then the situation is worse in a warming climate. If cold and snow were the limiting factors, then they might regenerate better in a new (warmer) climate.

[68] In area, compared to their proportional area in the mountains as a whole. I make no value judgments here (but see Chapter 9).

As high-elevation landscapes warm, often faster than the Earth as a whole (see Chapter 3), *energy limitation*, the principal limiting factor at treelines,[69] eases. We would therefore expect upper treelines to move higher, where terrain permits it. But there may be more fire in a warming West, and fires in the subalpine zone kill most trees that they reach. So on average, with more fire, we would expect the loss of trees at the highest elevations, lowering the upper treeline. Which of these opposite trends—trees' moving up because of climate or down because of fire—should we expect?

Probably both, and at a fine enough spatial scale that individual subalpine land-scapes will reflect the change.[70] In those subalpine landscapes that do or will experience fire—most but not all—one might imagine that some treelines will move higher, with climate superseding fire as a control, whereas others will move lower. Instead, we are more likely to see the changes over individual "views". That is, from a vista point on the landscape, the parkland will cover more elevation change and be more varied in its appearance, i.e., wider and more complex spatially. The reason is that more and larger fires, in contrast with longer growing seasons and less energy limitation, will compete as "forces of change" at multiple scales, from single trees and tree islands in complex topography to gentle slopes covering the elevation range of the parkland. The outcome will be a very visible change to more spatial variability and complexity[71] (Fig. 6.8).

Forests on the Brink? (Revisited)

We can say a bit more now about the *attribution* of forest dieback to climate change, a topic I touched on in Chapter 5. As I said, the state of knowledge is still uncertain enough that current researchers are circumspect about the attribution, even while documenting instances of dieback around the world. Based on our discussion of seedling vulnerability after disturbance, forest dieback induced by warming climate should have two elements: (1) massive fairly sudden mortality caused by a driver

---

[69] In general, over all life stages. In some arid mountain environments, it is actually desiccation that kills seedlings. This is a combination of water limitation and energy limitation, the latter in that cold prevents plants from accessing available water, either if it is frozen or their tissues cannot transport it.

[70] This discussion draws heavily on a recent paper (free online): C.A. Cansler, *et al.* 2018. Fire enhances the complexity of forest structure in alpine treeline ecotones. *Ecosphere* (2):e02091. https://doi.org/10.1002/ecs2.2091. This was an interesting case where the authors started out expecting one of the two "forces" to win out. Thomas Huxley said that "many a beautiful theory was killed by an ugly fact". But the facts, like these, can be more beautiful themselves, and still kill a theory.

[71] That visibility is a reason that this is a good example, but it is not the only one. The same interactions—warming climate, disturbance, succession—will play out in interesting and complex ways on many landscapes, such as closed-canopy forests, in which the changes may be harder to see from a vista point.

Fig. 6.8 A wide treeline in the northern Rocky Mountains, populated by two species with different adaptations to fire: subalpine fir (evades fire by arriving later in succession, and whitebark pine (colonizes sites after fire). Photo by Alina Cansler

strongly associated with climate change, such as a heat wave or unprecedented fire,[72] (2) failure of *recovery*, in this case meaning that the dead trees are not replaced by successful regeneration, i.e., of the killed species or another that "qualifies" as part of a forest.[73] This failure could be direct, where nothing comes back, or clear replacement by non-trees (Fig. 5.6).

"Climate-Change" Fires: Example (2) of Detection and Attribution

Are western wildfires in recent decades caused by climate change? As public attention on fires, firefighting expense, and obvious damage to human communities increase, the idea has gained more favor with the media and politicians. This

---

[72] OK I am intentionally "passing the buck" of attribution. How do we know that these events are "caused" by warming climate? More on that in Chapter 8. For now, it is simply a criterion.

[73] In some cases this is an obvious evaluation, in some not. For example, replacement of lodgepole pine by sagebrush would qualify, but what about replacement of pinyon pine by juniper, the big dieback events in the Southwest, or other low-stature pines by scrubby oaks, in California?

attribution problem is more difficult and technical, however, than the others I have discussed so far: glacier loss, sea-level rise, and forest dieback. Glacier loss and sea-level rise can call directly on our constant from Chapter 1, the melting point of ice.[74] Forest dieback is not so straightforward, but we have a logic of inference (just above, with our two elements) that we can follow, even though it is neither simple nor precise. With wildfires, for both technical and practical[75] reasons, we have to fall back on statistical procedures to identify the global-warming signal, if any, in the recent observational record of fires. Consequently, what we can say is limited, analogously to what we saw with global climate projections, in which the more precise we wanted our answer to be, the more uncertainty we had to accept. With very large fires, as with other *extreme events*[76] such as heat waves, we (ideally) back off from statements like "This wildfire was caused by climate change" to "Climate change makes fires this large and severe much more likely". Over time, given projected rates of warming, attribution will be clearer, for two reasons: (1) we will have a longer record, a larger sample; and (2) the differences between historical and (then) current fire seasons, in extent and severity, will increase, and as *detection* is clearer, so will be attribution. Even so, attribution of a specific wildfire will be problematic, and we will be safer with a *probabilistic* estimate of the likelihood of extreme fires, which is at least a better planning tool than the hindsight of attribution of fires that have already happened.

Landscape Disturbance Models: How Can We Use Them for Climate-Change Projections?

The models of interest to us here are similar in construction and use to the forest models from Chapter 5, but with added layers of disturbance. Recall that for a model to be of use for climate-change projections, some aspect of climate must be a driving variable, either a statistical or a biological type. I distinguish again between *forward* ("process-based") and *inverse* (empirical) models, both of which have different strengths and weaknesses and answer different questions.

---

[74] Even though not all sea-level rise is from melting snow and ice—thermal expansion causes a bit of it—the attribution is still quite clear. As ice on Greenland and Antarctica melts from global warming, sea level rises.

[75] The main technical problem is that the fires thought to be "climate-change" fires are rare. Both detection and attribution, of anything, are much easier when there are large samples. The statistics involved are what is called "extreme-value" theory, which compensates for the rarity of events with certain compromises that lead to more *uncertainty* than ordinary statistics. The main practical problem with fires is (curiously) the reverse of rarity: There are so many different circumstances and types of weather and vegetation that it is difficult to isolate the important factors in each case.

[76] More on these in Chapter 8. Their rarity is one difficulty, but not the only one.

We have seen how pervasive and important ecological disturbance is in the Western Mountains. So only now have we come finally to models that are comprehensive enough in conception[77] to address the future state of our mountain ecosystems. To succeed they have to "get right" the complexities of disturbance and succession in a non-stationary climate. The same general caveats apply as to the models in Chapters 4 and 5: sensitivity to initial conditions, changing outcomes at different scales, and non-stationarity. From a modeler's perspective, the most significant increase in difficulty is integrating the *stochastic*[78] nature of disturbance with the more "regular" processes of vegetation growth and succession. Recall from Chapter 3 that *ensembles* are the modelers' way of making lemonade out of the lemons of the stochastic nature of climate. With landscape disturbance models, the need for ensembles is now double, not only for the climate drivers but also for disturbances.

It might seem reasonable to approach disturbance models in two steps. First, convince yourself that you can use climate to predict disturbance. Then add some climate-disturbance routine to your simulation and from there simulate the effects on vegetation and other ecosystem properties. Let's use fire as an example of this process.[79] There has been some success in predicting the annual area burned[80] by wildfire in the West, as a response to climate. If we zoom out to the scale of *ecosections* (regions, generally smaller than states, which have "acceptably" homogeneous ecological characteristics such that ecological inferences can be made about the important drivers of fire), then empirical models for some of the ecosections are almost surprisingly good.

Step two in this example would be to use the empirical relationships from these models to simulate fire's effects on vegetation. The main drawbacks of this are two. First, annual area burned is a crude surrogate for what will actually happen to the vegetation, without knowing fire severity. Second, the models are at very coarse spatial scales, larger than individual fires, so any spatial distribution of fire on a simulated landscape will require some arbitrary choices. Nevertheless, this spatial scale more or less matches that of many *DGVMs* (see Chapter 5), obviating the

---

[77] Full disclosure: Unlike with other model types I have discussed, I have commissioned, written, reviewed, and used landscape disturbance models. I hope that the reader will see quickly that this does not make me blind to their limitations, even though I have just labeled them as the *sine qua non*.

[78] Meaning that they are different in each simulation. See Note 34, Chapter 3.

[79] Fire is the only example I will use for this first step. The empirical models for fire are the best developed for any disturbance, in the West. There are also models of insect outbreaks, but many of the issues are the same, so I won't cover them separately. For a good discussion of bark beetles and climate change, see B.J. Bentz, *et al.* 2010. Climate change and bark beetles of the Western United States and Canada: direct and indirect effects. *BioScience* 60:602–613.

[80] There are also models of "fire probability". This turns out to be a convoluted statistical transformation of area burned, but with enough numerical noise that the models are weaker than area-burned predictions.

problem of *scale mismatch*[81] that is a big part of our key pitfall of problems of changing scales. Curiously, however (in my view), most of the DGVMs that incorporate fire use algorithms that are meaningful at very fine spatial scales instead of matched to their own medium-scale resolution, creating a huge burden of false precision. There have been only a few cases in which area-burned models were integrated into climate change projections for ecosystems at the scale of ecosections.[82]

Most of the landscape disturbance models in use are forward models. Rather than the two-step process I described above, they build the forest (or non-forest) and fire components at similar scales from the beginning. For example, a forest model at 30-m resolution will be coupled with a fire-behavior and fire-spread model at the same resolution. Typically, the fire models will either approximate[83] the physics of combustion and fire propagation as directly as possible, or spread fire probabilistically. The latter turns out to be more practical for most landscapes of interest. For our purposes, a simple story line for landscape disturbance models is (1) Run everything in a forest model continuously for the period of interest (2) Simulate fires or insect outbreaks, and their effects on the forest, at a frequency and over an area that roughly matches what you think the average is, based on observations or some empirical estimate (3) Move those averages up or down based on how the disturbance is expected to change in a warming climate (4) Look at the results and see if they match observations where those are available and if they seem plausible where there are no observations (i.e., for the future). In short, landscape disturbance models for non-stationary climate are very much a work in progress. We have a reasonable idea of what a good one should be like,[84] but there are theoretical and technical problems in realizing these ideas. Some of the problems are not going away any time soon, because of intrinsic limitations that apply to most if not all ecological models.[85]

---

[81] Meaning that a process is being interpolated or extrapolated outside its range of validity. For example, if we ran a GCM at 1-km grid spacing (instead of 1°), there would be no way we could represent circulation meaningfully. Specifically, our outcomes would have *false precision*. In the case of fire, a recent paper modeled the association between climate and area burned at a very fine scale (~1°). The result was much obvious false precision, to be expected when climate and fire regimes were represented uniquely at that scale.

[82] Yes, we did that. C.L. Raymond and D. McKenzie. 2012. Carbon dynamics of forests in Washington, U.S: projections for the twenty-first century based on climate-driven changes in fire regime. *Ecological Applications* 22:1589–1611. The outcome was similar to that from the eco-hydrological models of Chapter 4. We asked what the difference was in carbon storage between present and future in Washington State.

[83] Why *approximate* physics? The true physics is understood at the scale of centimeters or even finer. This quickly becomes impractical on a computer when trying to cover even one 30 × 30 m area, taking longer by orders of magnitude to simulate combustion than it takes the same process to happen in real time.

[84] For a roadmap, as yet unrealized, see R.E. Keane, *et al.* 2015. Representing climate, disturbance, and vegetation interactions in landscape models. *Ecological Modelling* 309:33–47.

[85] For a review of these in landscape modeling, see E.A. Newman, *et al.* 2019. Scaling and complexity in landscape ecology. *Frontiers in Ecology and Evolution* doi:10.3389/fevo.2019.00293. (free online)

What Do We Expect for the Western Mountains?

Now that we have an idea of the full complexity of the forces of change in Western Mountain ecosystems, including hydroclimate, forest succession, and multiple disturbances, it should be clear that we are safer in staying with our first principles about what persists and what changes, rather than relying on model forecasts. For this chapter's "expectations", the ideas of *flammability-limited* vs. *fuel-limited* (Fig. 6.1) will be important throughout. The climate and hydroclimate models tell us a lot, the eco-hydrological models less, though with more specifics, and the landscape disturbance models even less with any certainty. Recalling that all ecological projections are about "what if" rather than "what will be", what do our first principles say about our mountain ranges specifically? I will focus on the major players: fire; insects, particularly MPB; and land movements. These are the disturbances about whose changes in a warming climate we have the most confidence. Table 6.1 summarizes the biggest expected players in each range.

Table 6.1 Changes in mountains expected to be most significant in a warming climate

Mountains	Hydroclimatic factor	Disturbance factor	Confidence
Cascade Range and Pacific Coast Ranges (N)	Maritime influence	Fire (E), Extreme fire (W)	High (E), Low (W)
Rocky Mountains (N)	Glacier loss	Fire and high-elevation MPB	High
Sierra Nevada	Snow loss	Fire/regeneration failure	High
Rocky Mountains (S and central)	Snow loss and drought	Bark beetles	High
Coast Ranges (S)	Drought	Fire/regeneration failure	Medium
Sky Islands and Basin Ranges	Drought	Drought-insects Fire/regeneration failure	Medium

How hydroclimatic factors ("Comments") from Table 3.1 interact with disturbance to drive ecosystem change. Confidence levels are the author's (I know of no consensus estimates for these confidence levels for disturbance in the Western Mountains, but for readers wondering how I determined them, there were two source of inspiration: (1) the IPCC's confidence estimates in their various assessments of the effects of climate change on the environment [https://www.ipcc.ch/report/ar5/wg1/] and (2) medical societies' (e.g., the American Society for Clinical Oncology [https://www.asco.org/] and the National Cancer Care Network [https://www.nccn.org/]) estimates of their confidence in their own recommendations for treatment protocols.)

The Cascade Range and the Pacific Coast Ranges (to the Klamath and Trinity Mountains)

The fire regimes in these ranges are much more predictable on the drier east slopes, and toward the southern ends. They are purely in the flammability-limited domain in the wetter areas, and on its edge in the drier areas, in that a big change toward warmer and drier could move the drier areas into being partly fuel-limited. In the near future, we can expect more wildfire area, and higher severity, on eastern slopes and in transition areas between there and the wetter western slopes. On the western slopes, with a strong maritime influence, most wildfires are associated with rare anomalous weather,[86] and the future in a warming climate is uncertain because it will depend on changes in global and regional circulation that may or may not occur.

In areas with abundant host species for MPB such as lodgepole or whitebark pine, warmer temperatures will accelerate the insect's life cycles. This will increase the potential for outbreaks at higher elevations, but may actually reduce it at lower elevations, in the range of lodgepole pine, because of the loss of *adaptive seasonality* that we discussed above. Given that there is much more area at lower elevations, we may well see less total mortality from MPB in these ranges in a warming climate.

With its rugged topography and abundant precipitation, the Cascade Range has more area disturbed by avalanches and landslides, on average, than by either fire or insect outbreaks. What I said above about the importance of hydroclimate for these disturbances applies to all the ranges of the Western Mountains. If winters are wetter and warmer in the Cascade Range, as predicted by regional climate models, slopes will be less stable, on average, but the potential complexities make predictions about more or less disturbance overall difficult. For example, the total precipitation includes both rain and snow, with the proportion of rain increasing in a warmer climate, reducing the total avalanche "burden". But avalanches depend not only on the total snowpack but also on its stability, which is a function of the variability in layers. So increased variability, for example from *rain-on-snow* events, may increase this burden.

The Northern Rocky Mountains

Most of this range is also *flammability-limited* (overlay Fig. 1.1 with Fig. 6.1), so with more warming projected than for the Pacific Northwest, we can expect an equivalently larger increase in wildfire area and severity. Some of our strongest models (statistically) are from Northern Rocky Mountain *ecosections*, so confidence in their projections of more fire is higher than elsewhere in the West. In the

---

[86] Specifically, strong (chinook) easterly winds, low humidity, and hot temperatures. Predicting these in association with global climate, in the Cascade Range and elsewhere such as Southern California, is difficult.

most water-limited ecosystems, however, as in the eastern slopes of the Cascade Range, even forests may become *fuel-limited*, thereby offsetting a range-wide trend toward more fire.

Large areas are dominated by lodgepole pine (e.g., Yellowstone), some of which[87] have already seen mortality from MPB outbreaks. These forests can be expected to continue, as long as there are enough forests with trees in vulnerable age classes. The prognosis for whitebark pine is worse in the Northern Rocky Mountains than elsewhere. It is more abundant, and more continuous there than in the Pacific Northwest, and it has already had significant mortality from MPB, whose dispersal into higher elevations is facilitated by warming climate.

The projected loss of snow and ice, worse here than elsewhere, may reduce land movements from their current level. They are smaller players than in the Cascade Range, and a smaller proportion of the total disturbed area, which is dominated by fire and insects.

The Sierra Nevada

These water-limited ecosystems are expected to become more so, with periodic droughts already taking their toll on the forests. Ecosystems could become more fuel-limited than they are now, if vegetation is sparser in response to drought. Fire is already very much a part of the landscapes, and may or may not increase with warmer temperatures, if fuel limitation increases. Probably the most severe effect of warming climate in the Sierra Nevada will be on post-fire succession. Already we are seeing forests burn and return as shrub fields (see Fig. 5.6). As marginal (too hot and dry) conditions for seedlings become more widespread, the number of these scenes will likely increase and move to higher elevations. Here and in other arid mountains, forests may indeed be on the brink.

The Pacific Coast Ranges (South of the Klamath and Trinity Mountains)

Recall the expectations for these ranges from Chapters 4 and 5. They are already near the edge of a climate that will support forests at higher elevations (woodlands, shrubs, and grasslands are at lower elevations). The post-fire environment, as in the Sierra Nevada, may discourage seedlings, whereas drought-tolerant species that can

---

[87] But not overly so in Yellowstone. It may just be a ticking time bomb, or have escaped the levels of damage further south (in Colorado) and north (in British Columbia, Canada) by having an active fire regime (thanks to mostly flat terrain and less fire exclusion than in other areas dominated by lodgepole pine) that keeps the proportion of the most vulnerable age class (80 years and older) down.

sprout from root stock after fire (one of our *evasive* strategies) may be fine. Either oaks or shrubs are likely to increase, even dominating the vegetation in some former conifer forests.[88] The fire regimes themselves may change, but given their dependence on the Santa Ana winds, differences in their global-warming future are as difficult to forecast as in the westside Cascade Range.

The Southern and Central Rocky Mountains

Like the Sierra Nevada, these mountains are mostly water-limited, and as I noted in Chapter 5 their complex pattern of landforms belies any true large-scale division into west and east slopes. The rain shadow is muted because the prevailing circulation brings air across the Great Basin. The biggest change in a disturbance regime will likely be increased insect outbreaks, much of which we are already seeing in Colorado, by the MPB and a relative, the spruce bark beetle. The latter's preferred host, Engelmann spruce, is a minor player in many forests where lodgepole pine dominates, but in the Central Rocky Mountains it can form almost pure stands that are therefore susceptible to massive mortality[89] in outbreaks.

The Sky Islands and the Basin Ranges

Our expectation (see Chapters 4 and 5) that these ranges will be hardest hit of all by the changes in hydroclimate holds when we add disturbance to the mix. For example, a combination of drought and insect attack (the *Ips* pinyon beetle) has created massive mortality in pinyon pine in the Southwest. Moreover, the same post-fire issues with regeneration should be expected as in other arid mountains. Forests are thus being hit from several directions: mature trees stressed by drought are more vulnerable to disturbance, and seeds of trees killed by disturbance are more vulnerable to early mortality from drought.

---

[88] In the Western Mountains, populations of oaks are rarely called forests, rather woodlands, if large trees such as the oaks of the Pacific Coast Ranges, or shrublands, if the smaller oaks inland. This is partly subjective, and can be misleading in areas like carbon accounting, especially because the large-oak woodlands in the Coast Ranges have a lot of biomass.

[89] Remember that we need a critical mass of resources, whether food (for defoliators) or tissue for cambium feeders and their galleries, to enable and sustain an outbreak.

Chapter 7
Creatures Great and Small

All things bright and beautiful
All creatures great and small
—*Anglican hymn lyrics by C. Frances Alexander*

It is better to be a tiger for one day than a sheep for a thousand
years
—*Tibetan proverb*

The first members of the animal *kingdom*[1] appeared on Earth sometime after the first multicellular organisms, thought to have originated around 800 million years ago, but the largest pulse ever of biological diversity in the animal kingdom happened during the "Cambrian explosion", which began around 540 million years ago, and lasted less than 25 million years. During that "brief" *evolutionary time*, all the modern *phyla*[2] appeared, along with some that are now extinct. Since then, the animal kingdom has seen some severe bottlenecks. The worst of these was at the end of the Permian era, around 250 million years ago, when about 70% of all terrestrial vertebrates were lost, along with about 95% of all marine species, including the iconic trilobites, which survived several other major extinctions. The best known mass extinction is of course the one that was the demise of the dinosaurs, around

---

[1] The broadest division in the biological taxonomic hierarchy of use for us. Following the most recent classification system, by Ruggiero, Cavalier-Smith, and colleagues in 2015, there are seven *kingdoms*: Bacteria, Archaea, Protozoa, Chromista, Plantae (plants), Fungi, Animalia (animals). In this book, plants and animals are the focus. See Chapter 5, Note 6, regarding other levels. M.A. Ruggiero, *et al.* A higher-level classification of all living organisms. *PLOS ONE.* 10 (4): e0119248.

[2] The next division below kingdom. The animal phyla of most interest for us are *Arthropoda*, with over a million land species, including insects, and *Chordata*, mainly the subphylum Vertebrata, including mammals, birds, reptiles, amphibians, and fish, with about 70,000 species.

© Springer Nature Switzerland AG 2020
D. McKenzie, *Mountains in the Greenhouse*,
https://doi.org/10.1007/978-3-030-42432-9_7

66 million years ago, at the *Cretaceous-Paleogene*[3] boundary. The importance of this last major extinction for animals of the Western Mountains is twofold: the survivors of the dinosaur lineage evolved into the birds of today, and the ecological "space" vacated by the dinosaurs provided opportunities for mammals, already on Earth, but small, not numerous, and lower on the food chain. This is one visible way in which evolution changes course.[4]

The great glaciations of the *Pleistocene* epoch were a bottleneck for North American animals, as they were for trees and other plant species. Even though the ice ages were not extinction events *per se*, they constricted movement, both dispersal and migrations, such as to increase the evolutionary isolation of the Western Hemisphere. The animals of the Western Mountains are a legacy of this isolation, but they are only a subset of what existed at the beginning of the *Holocene* epoch.[5] The largest mammals were disproportionate victims in North America[6]; all disappeared between the end of the Ice Ages and the Euro-American conquest of the West. Various theories exist, including human hunting, climate change, and diseases, but it is perhaps more than chance that most *megafaunal* (literally, big animals) extinctions world-wide in the last 50,000 years seem to have coincided with the first arrival of humans to the area. In North America, this included mastodons, saber-toothed cats, giant ground sloths, and the giant short-faced bear.[7] The large mammals of the Western Mountains today are the survivors.

Historical Effects of Humans on Western Wildlife

Whatever the cause of the Holocene megafaunal extinctions, the arrival of Euro-Americans has reduced further the proportion of large mammals among the wild species remaining in the West. Although few species have actually become extinct,

---

[3] This has been known for years as the "Cretaceous-Tertiary", or "K-T" event. For those interested in the details, the new name reflects a practice of naming both sides of a boundary with the same level in a classification. The Cretaceous was the last "period" in the Mesozoic "Era", and the Paleogene (not to be confused with the "Paleocene", which is an "epoch", one of seven) the first period of three in the Cenozoic Era, in which we are today. For reference, Earth's history is divided, in order of finer partitions, into (1) eon, (2) era, (3) period, and (4) epoch. See https://en.wikipedia.org/wiki/Geological_history_of_Earth for details.

[4] Of many. The idea that evolution moves mostly in stops and starts is an alternative to continuous change and has a fair amount of evidence behind it. This topic is outside our scope, but see S.J. Gould, and N. Eldredge. 1977. Punctuated equilibria: the tempo and mode of evolution reconsidered. *Paleobiology* 3:115–151.

[5] Recall that this is the period after the last glaciation ended, beginning roughly 12,000 years BP.

[6] More so than elsewhere, apparently. I have seen a second-hand report of a study that confirmed this, but it is in an obscure and unavailable book chapter.

[7] For details, see J.T. Faith and T.A. Surovell. 2009. (free online) Synchronous extinction of North America's Pleistocene mammals. *Proceedings of the National Academy of Sciences, USA.* 106:20641–20645.

local *extirpations* (loss of a species in some defined portion of its total range) have been widespread. This is a major difference from outcomes for trees[8] and other plants. Shrinkage in *distribution*, or range,[9] has been severe for vulnerable species such as grizzly bears and wolves, to moderate for more adaptable species such as mountain lions, to almost none for the resilient coyote. In the case of the American bison, a herd animal, population density, i.e., the size of herds, has shrunk along with its distribution.

Extirpations have come through two mechanisms. The direct one is active hunting. In conjunction with the movement of Euro-American people west, and well into the twentieth century, the killing of predators was subsidized, with bounties on bears, lions, wolves, and coyotes. For example, the last grizzly bear in California, whose state flag has its likeness, was killed in 1922. The second mechanism is habitat loss, usually to development, often preceded by resource extraction such as clear-cut logging. This either deprives animals of critical resources or brings them into fatal contact with humans.

These "top-heavy" losses mean that the baseline[10] for anticipating the effects of warming climate on animals of the Western Mountains is not what it would have been had not humans removed certain species selectively.[11] Climate controls animal species' distributions, as it does plants', but for animals that have been subject to heavy losses from human activity, the *realized*[12] distributions are often a small fraction of what they would be if climate and other non-human ecological forces were the only ones acting. Similarly, the interactions that limit species distributions beyond just what is determined by climate, such as inter-species competition for resources, can be very different if populations are reduced or extirpated. Some species are very important for ecosystems in this regard, such that many processes and patterns in ecosystems will change with their absence or scarcity.

---

[8] Yes, we have lost forest acreage, and in particular old-growth forest, which has been replaced by either young stands that have regenerated naturally, single-species plantations, or shrublands. But species, trees in particular, are mostly still where they were 200 years ago.

[9] Distribution refers here to where a species is found, as opposed to an individual animal, which has a *home range*. See text below.

[10] As we've already seen, of course, there is no unchanging baseline, but we are farther from any equilibrium than we would have been without the human influence.

[11] And on a massive scale, unlike the selective use of animal and plant species by Native Americans.

[12] You may have encountered the phrases "realized niche" and "fundamental niche", referring to the difference between where a species is observed and where it could be if climate were the only control.

Keystone Species

In the parlance of conservation, *keystone* species are those that "hold things together" from above.[13] In an ecosystem, it may be a top predator, such as bear, cougar, wolf, or eagle in the Western Mountains. For example, the loss of wolves from the *Greater Yellowstone Ecosystem* to extermination by Euro-Americans removed the principal control on *ungulate* (hoofed mammals, in this case elk and deer) populations, which in turn increased browsing on their favorite food, quaking aspen. The result was a major change in the composition of tree species, with further implications for their "users".

As the most visible creatures in an ecosystem, keystone species are often the first to be noticed when their numbers change. They are also often the most vulnerable to extirpations because of that visibility. Their vulnerability to a warming climate varies, however, depending on their *life history* traits. These are accumulated "moves" in the evolutionary game, made by species in response to their environments. Because animals are *vagile*, these are mostly different concepts from those that apply to plants, but the general notion is the same.

Home Range, Migration, Dispersal

Life is certainly different for organisms that can move than for those that cannot. In general, plants have adapted to the changes that occur, daily, seasonally, and annually, in one place, whereas animals have adapted also to spatial patterns on landscapes[14] that may be *homogeneous* or *heterogeneous*.[15] An animal's spatial universe is its *home range*. This refers to all the locations it visits regularly, with whatever frequency, when not *migrating*. Migrations, which I distinguished from dispersal

---

[13] As opposed to species, or even kingdoms or phyla, that are essential players at the bottom of the food chain, such as bacteria, algae, or the mycorrhizal fungi that we saw in Chapter 5. There is much to be learned about how these and their interactions will change in a warming climate, but I am leaving that topic to others.

[14] These landscapes can be tiny enough to be thought of as "one place", for example a log in which a burrowing insect spends its life, or up to hundreds of square miles.

[15] It's important to understand that which of these a landscape is depends on the perspective of the "user". This is partly due to the scale of interest for the user, and partly due to the "use" of the landscape. For example, for a human, a mosaic of forest and meadow, perceived from a vista point above, seems heterogeneous as a whole; for a species using only the meadow or the forest, whether plant or animal, that subset of the landscape is homogeneous. To the second point, a *raptor* (a bird that hunts and feeds on other animals) may perceive an open-canopy forest as a homogeneous prey source, whereas its prey may see the gaps and the closed areas as very different, i.e., unsafe vs. safe.

for plants in Chapter 5, are seasonal—out and back—or once in a lifetime[16] moves to a different home range. *Dispersal*, as with tree species, is a permanent move out of the current home range, but unlike with plants, whose seeds and pollen disperse, animals disperse during their own lifetimes. For many mammals, particularly carnivores that hunt in groups, this happens as juveniles, when the social structure of the group does not permit juveniles to stay with the adults. The MPB that we saw in Chapter 6 as an important biotic disturbance agent thus has a fish-like dispersal: it disperses during its lifetime to a vulnerable host tree to lay eggs in the cambium.

An important distinction between migration and dispersal is that migration is more or less hard-wired; annual migrations are to a fixed location[17] (although with a large home range that can be a large area), whereas dispersal is generally opportunistic. So migrating animals move, for example, to a winter or summer (alternate home) range; dispersing animals move and establish a new home range.

Vulnerability to a warming climate can be due to changes in any or all of home range, dispersal, or migration. Loss of quality or quantity, or both, of home range, whether the single home range of a non-migrating species or one of the home ranges of a migrating species, reduces a species' *viability*.[18] If a migration route is compromised, then only some or none of a migratory group can reach its alternate home range, and may use too many resources to do so. Similarly, if a dispersal route is compromised or there is no place to establish a new home range, those animals are stressed and may not survive. Sometimes both home-range existence and travel are compromised, whether due to climate change, or human activity, or both. We shall see some examples below, but first we need a few more concepts in order to understand the future prospects for animals of the Western Mountains.

Habitat Quality, Refugia, and Connectivity

Habitat is simply the three-dimensional space in which an animal lives. *Core habitat* is the least dispensible part of its home range, providing resources without which it cannot survive, and usually located in the interior of its home range. *Peripheral* habitat is more replaceable. For example, core habitat for a bird can be trees suitable for nests, whereas peripheral habitat may be where it forages intermittently. In this

---

[16] In the Western Mountains, once-in-a-lifetime moves are by *anadromous* fish, those who spend their adult lives in the sea and migrate into fresh water, i.e., western rivers that empty into the Pacific Ocean to spawn, and then die. For those who read about this topic, you may see the term *semelparous* referring to salmon and other fish species, meaning reproducing only once, as opposed to *iteroparous*. (and the opposite of "anadromous" is actually "catadromous", meaning those that live in fresh water and migrate to salt water to spawn).

[17] For example, anadromous fish *migrate* (once) back to the stream of their birth. This is one case where a migration is also a *dispersal*.

[18] This is the widely used conservation jargon for the chances that a species can avoid local or global extirpation. The latter is of course equivalent to extinction.

case, core habitat is clearly defined, but in other cases, such as with an animal that uses different parts of a large home range equally, the concept is less meaningful. Habitat quality refers to how well habitat serves the animal's needs.[19] Low-quality habitat generally means lower fitness, in the evolutionary sense. Sometimes quantity can substitute for quality, if the quantity is available. For example, home ranges are typically larger for animals of the same species where the habitat quality is lower, but only if there is area that is "unclaimed".[20]

Refugia, in ecology, are areas[21] that shelter one or more populations of plants or animals in an otherwise hostile landscape, i.e., non-habitat. Some have been created administratively, such as the historic Arctic National Wildlife Refuge, whereas others, usually unnamed, emerge from ecological processes that "miss" some places on the larger landscape. For example, on fire-prone landscapes some areas seem never to have burned, or to have burned much less frequently than the average. *Riparian* (next to flowing water) zones and wetlands can do this for obvious reasons, but sometimes the reasons fire refugia exist are harder to identify, such as when topography modifies the passage of air through an area such that a fire would have to travel upwind to burn it. As one might expect, refugia are often isolated, and make up a small percentage of the landscape.

Connectivity refers to how easy it is to travel from one place to another. This is an issue for *non-volant* species (those that don't fly) in the Western Mountains, in which terrain is often steep, rugged, and varying. When invoking this concept we must always keep in mind "connectivity for whom?". The most obvious contrast is between aquatic[22] and terrestrial species; for the former, river and stream networks are connected, everything else is not. For the latter, species differ greatly in their ability to navigate what is often difficult terrain, whether because of topography, or climate, or both. For example, the Sky Islands (see Chapter 2) are not connected for their non-volant[23] mammal inhabitants. Even though they are all capable of walking on the flat ground of the deserts in between, they would not survive the heat and dryness of the climate.

A key component of connectivity is its fragility. If there is only one corridor for travel in an otherwise difficult landscape to cross, then connectivity is very fragile. The classic case is a dammed river with no ladders or other means for fish to pass. On land, this can be a strip of forest, such as along a stream, on otherwise treeless terrain, providing protection from aerial predators, shelter, or moderation of the climate. It can also be the opposite: a passable corridor for a large animal in other-

---

[19] It may be useful to think of this as value per acre, or some other unit of area, similar to the value of real estate economically. Just as a human can get more cash (e.g., property rental) from higher-quality "habitat", so can an animal extract more of an essential resource from a smaller area if the quality is high.

[20] Either by another animal of the same species or if the area is bounded by non-habitat.

[21] Literally *refuges*, but this term is used administratively for areas with a specific purpose, such as a wildlife refuge. Perhaps that is why the Latin term *refugia* (singular *refugium*) was adopted.

[22] The non-volant ones, i.e., excluding aquatic birds.

[23] And probably not for their bats either.

wise very dense forest. We sometime measure the overall connectivity of a land-scape with a metric called *landscape resistance*, referring to how much the landscape overall impedes travel.[24] Changes in landscape resistance in response to some exter-nal action, such as a dam or a patch of clearcut logging, measure the aforementioned fragility of its connectivity. This may seem like a lot of jargon; the main point is that we want to measure things on the broad scales at which we can detect changes in climate, and want to have a view from a moderate distance. That often requires numbers and averages. Of course, in some cases, like the dam, we know what is going on without calculating a geeky metric at all, but most situations are not so clear.

Generalists and Specialists

We distinguish generalists vs. specialists in the animal world by looking at diet and habitat. Like humans, other animals can be *herbivores* (vegetarians), *carnivores* (meat-eaters), or *omnivores* (no restrictions). Carnivores have scientific sub-categories for fish-eaters (piscivores) and insect-eaters (insectivores). So broadly speaking, omnivores are generalists and the other categories are specialists, with gradations. Many species have a primary source of food, even if it is not the only one. A good functional definition of a diet specialist is an animal that is dependent on one food source.

Habitat specialists rely on a specific habitat type to survive. For example, among birds, some are bound to old-growth forests (e.g., northern spotted owl), whereas others thrive in seemingly "dead" forests after high-severity fires (e.g., black-backed woodpecker). Notable habitat specialists among birds are the *cavity-nesters* (build-ing nests in sheltered spaces, typically in tree trunks), some of which, the wood-peckers, create their own cavities. Among mammals in the Western Mountains, notable forest specialists are many[25] of the so-called *meso-predators* (literally, medium-sized species that hunt others), such as the American marten and the fisher. Our larger mammals at the top of the food chain, such as the grizzly bear, gray wolf, and mountain lion, are generalists; they can thrive in many habitat types.[26] All other things being equal,[27] habitat generalists would be less vulnerable to extirpations than are specialists.

---

[24] Recall again that this is very species-dependent.

[25] But then some, such as the raccoon, are very much the opposite.

[26] Or could, before the concerted efforts of humans to eliminate them from the West.

[27] Which of course they're not. See Note 26.

Landscape Changes that Matter

What could be bad news for animal species in a warming West, and which ones will be affected? We look below at some example species, with different vulnerabilities to a warming climate, other environmental changes, and their interactions, but first we cover some global issues.

Loss of Refugia

Most refugia are isolated patches of a habitat type that covers larger areas elsewhere. For example, in the Pacific Northwest east of the crest of the Cascade Range, patches of old-growth forest have literally "admitted" refugees—northern spotted owls—from the liquidation of the old growth forests west of the crest. If these refugia were to burn in higher-severity fires associated with a hotter drier climate, it would be a critical habitat loss for the owls. In general, refugia are not stable enough in *evolutionary time* for species[28] to evolve within them. They are actually most important for species that have colonized them from elsewhere when under pressure to disperse, so their loss can be (the last) part of a sequence of unfavorable events for a species.

Widespread Loss of a Type of Vegetation

The Western Mountains have enough habitat specialists that regional loss of any major vegetation type—forest, meadow, sagebrush—will affect multiple species. Some examples of losses include the forest dieback that I introduced in Chapters 5 and 6, replacement of sagebrush or other shrubland by annual grasses,[29] and loss of alpine meadows to an advancing upper treeline.

Rising Stream Temperatures

Aquatic species are generally adapted to a narrower range of water temperatures than are terrestrial species to ranges of air temperatures. Periods of elevated temperatures—the equivalent of water heat waves—are therefore more lethal to fish and

---

[28] Excluding micro-organisms that evolve in for what larger species is *ecological* time. See discussion in Chapter 1.

[29] The worst cases of this are when the grasses are *invasive species* (literally, coming from elsewhere, whether from nearby or halfway around the world) such as cheatgrass and buffelgrass, both of which "fill in" the spaces between sagebrush plants and make the whole ecosystem more vulnerable to large fires.

other aquatic species than the same temperature increases are to most land animals.[30] Given the limited connectivity of stream networks, especially in the south-north direction (i.e., to cooler climates), many fish species are vulnerable to the increased stream temperatures that will accompany a warming climate.[31] In some cases, there could be a double effect from reduced levels of dissolved oxygen associated with warmer water temperatures.

Changing Patch Structures

The species with larger home ranges, typically raptors and top predators among mammals, cross different *patch types*[32] in moving through their range or dispersing. Several aspects of these patches are important for the animals that cross or occupy them: different aspects at different values for different species and different ages and sexes within species. Habitat quality, the distribution[33] of patch sizes and shapes, and connectivity between patches of different habitat quality all make a difference. Neighborhood relationships also matter. For example, are all the patches of high quality near each other, or dispersed (like a checkerboard)? Animal species vary in how important any or all of these aspects are to survival and reproductive success.[34]

---

[30] For example, we saw that the MPB, which as an insect is *pokilothermal* ("cold-blooded", as opposed to *homeothermal*, or "warm-blooded") adapts to temperature changes by modifying its metabolism and therefore its reproductive cycles.

[31] Because of the limited tolerance of aquatic species, this is a problem even though water temperatures do not rise one-to-one with air temperatures. The "flattening" is due to the evaporative cooling that we discussed in the context of forests and hydroclimate. For details on this, see O. Mohseni and H.G. Stefan, 1999. Stream temperature/air temperature relationship: a physical interpretation. *Journal of Hydrology* 218:128–141.

[32] I used this term earlier informally (Chapter 6) to refer to groups of trees, but its principal use is as landscape-ecology jargon to refer to areas of relatively homogeneous vegetation (or the lack of it, e.g., on rock, snow, or ice). A classic simple spatial pattern of patches is a checkerboard, with identical black and red square "patches", 32 of each.

[33] For example, are all patches the same size, and is that large or small? Are there a few large ones and many small ones, or more or less equal numbers of each size? Are they close to round, meaning there is maximum patch interior for the amount of edge, or the opposite, with irregular perimeters and minimal interior?

[34] A full treatment of spatial patterns on landscapes is beyond our scope. The science of landscape ecology addresses those and how they relate to ecological process. For a deep but rewarding dive into that subject, see M.G. Turner, *et al.* 2015. *Landscape Ecology in Theory and Practice*. 2nd Edition, Springer Books.

Some Examples

Climate change rarely acts alone on animal species in complex environments. Remembering from Chapter 4 that the ongoing and expected losses of glaciers are the purest example in the Western Mountains of the effects of warming climate, we can look at different vulnerable species to see how much of their "trouble" is due to the warming climate as opposed to other factors.

Mostly Climate

The Canada lynx is a mid-sized cat (it can weigh from 11–40 lbs.), about the size of a bobcat and twice the size of a domestic cat (Fig. 7.1). Its current habitat extends far north into its namesake, but its southernmost ranges extend only a short way[35] into the northern USA. The lynx is a carnivore and diet *specialist*, with the snowshoe hare being its preferred prey. The hare is a swift runner, but the lynx's paws are exquisitely adapted to running on snow, making it a much more successful hunter

Fig. 7.1 Canada lynx. Photo from pixbay: https://pixabay.com/

---

[35] The northern parts of Washington, the Lake States, and New England.

on a surface of snow than on any other. Canada lynx home ranges vary in size, depending on habitat quality and the density of snowshoe hares. Although they cross different types of terrain in hunting or dispersing, their *obligate* habitat type (needing that habitat specifically) is open woodlands that provide enough cover to approach their prey but are not so dense that they are slowed down in pursuit. So in that sense they are also habitat specialists.

In a warming West, the US population of Canada lynx will be endangered. Projections of climate suggest that snow cover will be reduced greatly, depriving the cat of much of its prime hunting ground, and also reducing the number of snowshoe hares. A secondary ill effect of warming climate may be more wildfires within its core habitat in Washington State. Humans constrain Canada lynx in its southern range, but this is definitely secondary to climate.[36]

The wolverine (Fig. 7.2) is widely distributed in the boreal regions, but like the Canada lynx, its range extends only a bit into the northern USA, and only in the Western Mountains. Unlike the lynx, the wolverine is mainly a scavenger, making it a diet *generalist*. Wolverines can have very large home ranges, up to hundreds of square miles, but they are still habitat specialists in one important way: female wolverines burrow into the snow in mid-winter to create a den, and these dens are used

Fig. 7.2 Wolverine. Photo by Jeff Copeland

---

[36]We studied this intensively a few years ago, but unfortunately no paper is unpublished yet.

well into spring. This means that high-quality habitat for the wolverine requires snow cover until late spring.

In a warming West, the US population of wolverine will also be endangered, particularly in the northern Rocky Mountains in which temperature increases are projected to be greater than in the Cascade Range. As with the lynx, wolverine habitat is compromised by humans, mainly through hunting and trapping, but these have already had their effects. The loss of snow cover, the most direct environmental consequence of a warming climate, is now the principal threat.

Mostly Habitat Loss

The grizzly bear is clearly a top predator, but also an omnivore. It preys on large animals when available, fish, and birds and their eggs. It is also a scavenger and herbivore. Before the extirpations of the nineteenth century, the bear inhabited the entire West, even ranging into the Plains states, but by 1970 it was absent from about 98% of its historical range. Its extirpations were due both to active hunting, subsidized by the federal and state agencies, and to habitat loss from land uses that included urban development and agriculture. The historical range demonstrates that the grizzly is very much a habitat generalist, and that a warming climate *per se* will not be the threat that it is to Canada lynx and wolverine.[37] Were it not for human-caused barriers to dispersal, the bear could expand into at least part of its former territory. Given those barriers, however, and the expected stresses on ecosystems and societies in the warming climate, we can expect the bear to be increasingly vulnerable to extirpation in the Western Mountains.[38]

The gray wolf (Fig. 7.3) is also a top predator, and a pure carnivore, hunting in packs that can bring down most any animal. Like the grizzly bear, it has been subjected to aggressive subsidized hunting, such that its range in the United States has shrunk from two-thirds of the country to areas of the northern Lake States and parts of Montana, Idaho, Oregon, Washington, and around Yellowstone in Wyoming. So also like the grizzly bear, the gray wolf is a habitat generalist and it could expand into its former range even in a warming climate if human-caused barriers were absent. Being smaller and more fecund[39] than the grizzly bear, it is probably less

---

[37] For example, if we observe that current grizzly-bear habitat in the Western Mountains is in their far north—the Northern Rocky Mountains and bits of northern Washington and Idaho—we can infer that if these areas warm, they will become more like the bear's former southern habitats, such as the mountains of California and the southern Rocky Mountains. It was not the warm climate there that eliminated the bears, but humans.

[38] For example, future drought and forest loss may reduce both the arable land base and the viability of protected areas (wilderness and national parks). More on this in Chapter 9.

[39] Wolves form mated pairs, for life (of the shorter-lived member), and are sexually mature by 3 years, with pairs having 1 litter per year beginning sometime after that, depending on available resources. Grizzly bears are not sexually mature until 5 years, and females have an average of two cubs every other year.

Fig. 7.3 Gray wolf, part of a population that recolonized northern Washington State. Photo courtesy of Conservation Northwest. https://www.conservationnw.org/

vulnerable to the aforementioned stresses, but still faces an uncertain future in the midst of society's responses to warming climate.

A third top predator and a carnivore, the mountain lion (Fig. 7.4), is not vulnerable to West-wide extinction, but because lions are habitat generalists, populations persist in unlikely places, such as the Santa Monica and Santa Ana mountains of southern California. They have persisted as development has proceeded all around them, but even they cannot cross hard barriers to dispersal such as the notorious freeways of the Los Angeles Basin. Consequently, these populations are isolated and are inbreeding (even adult lions with their offspring), and without new genetic material (from assisted dispersal) will likely be extirpated in the next few decades.

Habitat Loss and Climate

The northern spotted owl (NSO) is well known for its key role in the ecology and politics of conservation. It is a carnivore, feeding mostly on small mammals but also reptiles, other birds, and insects. Although it is debatable whether the NSO is a true

Fig. 7.4 Mountain lion in the Northern Rocky Mountains, where it is not threatened (see text). Photo courtesy of People and Carnivores. https://peopleandcarnivores.org/

habitat specialist,[40] it is definitely intolerant of habitat disturbance. In Washington State, for example, it is estimated[41] that the 90% loss of old-growth forest to logging has caused a decline of 40–90% in NSO numbers. In the past, habitat loss was the principal cause of this species' vulnerability, but in the future, it could change to climate and its associated disturbances such as more wildfire in habitat refugia, whether west or east of the Cascade-Range crest. A further threat, unrelated to warming climate *per se*, is the westward dispersal over the last century of the barred owl from its historical range east of the Mississippi River. Barred owls displace NSOs by disrupting their nesting and competing with them for food, such that in much of the NSO's range they are outnumbered by barred owls. So the NSO faces multiple threats, any or all of which could be lethal.

---

[40] In the northern part of its range, Canada to southern Oregon, the NSO historically lived only in old-growth forests, whereas in the southern part, the Pacific Coast ranges of California, it is found in a mix of old and younger forests. With the extensive loss (more than 90%) of old-growth forests in western Washington and Oregon to logging, some owls dispersed across the Cascade-Range crest to the east, nesting (apparently) successfully in *refugia* of older forest. Whether these populations are stable depends on many things, including climate change and changes in fire regime (see Chapter 6), and has caused considerable political heat.

[41] Depending on who is estimating, the numbers change, thus the wide range that I give.

Fig. 7.5 Over-wintering monarch butterflies at San Jose Creek, near Goleta, CA. These are from the western population. Photo by David James

The monarch butterfly is widely known for its seemingly impossible migrations. There are two distinct populations. The better known of these spends summers in the eastern United States and over-winters in the Angangueo Mountains of central Mexico, in the Monarch Butterfly Biosphere Reserve. The western population spends summers in the West, and over-winters in southern coastal California (Fig. 7.5). Much of what we know about threats to the monarch comes from studies of the eastern population, which has lost habitat in its summer range, *stepping stones* (literally, places to stop and "refuel", reproduce, or both) on its long migration, and some of the formerly intact forest in its over-wintering area.[42] Both populations are challenged[43] by climate change and loss of habitat or food source. Over-wintering grounds for the western population are dispersed across southern

---

[42] Even though the Reserve is protected, the protection is not the same as that in a U.S. wilderness area or national park. The local population has both legal and surreptitious access to the forest; some logging is permitted and some trees are poached.

[43] Our focus on the West means that this example describes challenges to the western population. For good science on the eastern population, search the internet for work by Karen Oberhauser on the effects of climate, or Lincoln Brower on the biology, or most any logical keywords. Many papers are free online. The most severe threat to the over-wintering population is extreme weather (i.e., more than loss of forest cover at current levels), which is exacerbated by sparse canopy cover. For the migrating population, loss of its chief food source, milkweed, may be the most severe.

coastal California, as opposed to concentrated in a reserve. This is good and bad news; they are less vulnerable to a single event, whether a storm or a large logging operation, but not officially protected. Their breeding habitat is closer, on average, to their over-wintering grounds than that for the eastern population, so the loss of stepping stones is not quite so lethal. The most recent research suggests that both land-use change and climate are significant as threats to the viability of the western population.[44]

The bull trout is a cold-water fish, a salmonid,[45] that was once abundant in streams and lakes of western North America. They now occupy only about 20% of their historical range, thanks to a combination of human-caused habitat loss and climate-related deterioration of once ideal streams. In the language of fish biologists, bull trout require the "Four Cs": habitat that is Cold, Clean, Complex, and Connected.[46] The main climate-related threat is rising stream temperatures across the West ("Cold"), but the "Complex"—streams with riffles and deep pools, undercut banks and lots of large logs—is also vulnerable to increasing variation of seasonal and peak flows. The "Clean" (water quality) and "Connected" (no dams or other barriers to their dispersal), are human-caused.

Other salmonids, as *anadromous* fish (see Note 16), migrate for many miles to reproduce at the ends of their lives, reversing the journey that they made in their youth. So they are sensitive to loss of habitat at both ends of their migration, and to barriers, mostly human-created, along the way. Ongoing declines in *wild populations* (as opposed to those grown in hatcheries and released), for example the roughly 90% decline from historical numbers in Chinook salmon in the Pacific Northwest, have come from reduced habitat quality (in the case of Chinook, in the Puget Sound) and the only partial effectiveness of artificial aids[47] to migration. Moving into the future, however, a climatic factor will become more important. As I mentioned above, most aquatic species have fairly narrow ranges of water temperature that they can tolerate. For the West's anadromous fish, temperatures in many streams are reaching the upper ends of these ranges.

---

[44] E.E. Crone, *et al.* 2019. Why are monarch butterflies declining in the West? Understanding the importance of multiple correlated drivers. *Ecological Applications* 9:e01975. https://doi.org/10.1002/eap.1975.

[45] Taxonomically, salmonids are both a family (Salmonidae) and an order (Salmoniformes). See Chapter 5, Note 6. They include salmon and trout, and are primarily *anadromous*. They are all predators, eating small crustaceans, aquatic insects, and smaller fish.

[46] This information, with the "Four Cs", comes from the US Fish and Wildlife Service's web pages. https://www.fws.gov/pacific/bulltrout/.

[47] For example, fish ladders over dams in formerly wild rivers of the Pacific Northwest.

The Most Sensitive Groups: Amphibians and Reptiles

In general, mammals, birds, fish, and insects (and plants) vary in their sensitivity to warming climate, depending on *life history* and whether they are habitat generalists or specialists. In contrast, we believe[48] that in the Western Mountains, many or even most amphibians and reptiles are vulnerable. This is due to physiological limits, particularly their being *pokilothermal* (see Note 31), and to their limited ability to disperse, particularly amphibians, many of whose populations are already decimated by a lethal fungus.[49]

Some Winners

We saw in Chapter 6 how the MPB thrives on warm temperatures in that its metabolic rate increases to permit faster reproduction. Other insects thrive similarly, with negative consequences for larger organisms. For example, both mosquitoes and the viruses they carry develop more quickly in warmer temperatures, and mosquito bite rates increase. This may already be moving the latitudinal limits away from the Equator for diseases like malaria and dengue fever, and making West Nile virus infections,[50] which are already affecting the Western Mountains, more common. Another winner may be the deermouse, ubiquitous in the West and the carrier of the deadly hantavirus, of which more cases appear in humans in warmer drier years.

A strong survivor, if not an outright winner, will be the wily coyote. A habitat generalist, and a hunter and scavenger and a definite omnivore (including fruits and plants), the coyote has persisted through the extirpations of the grizzly bear and gray wolf, and not for lack of efforts to exterminate it. Even though a half-million coyotes are killed every year, many gunned down from the air, the coyote has actually expanded its territory, originally west of the Rocky Mountains, to the East Coast, and expanded its habitat into cities, to the chagrin of many a careless household cat. One remarkable adaptation of the coyote is a flexible breeding strategy, something they share with jackals. Under pressure packs split up and coyotes form breeding pairs. They also adapt litter size to population pressure[51]; it ranges from 5–6 pups with no population pressure to 12–16 where there is.

---

[48] For those readers who like graphics and data, I recommend M.J. Case, *et al.* 2014. Relative sensitivity to climate change of species in northwestern North America. *Biological Conservation* 187:127–133. Their assessment is a brute-force survey of experts on 195 species of plants and animals.

[49] *Chytridiomycosis*, literally fungal ("myco") attack by a *chytrid* (fungus), has no known direct association with global climate change, but like habitat loss, it compounds the vulnerability of species already compromised by rising temperatures.

[50] Less than 1% of humans who are affected develop severe neurological symptoms, and about 10% of those die. Horses are also vulnerable to serious disease from the virus; before a vaccine was developed about 40% of those infected died.

[51] How does a wide-ranging pack animal manage the population census needed to "decide" to adjust breeding strategy? From how often a pack's howls are answered by another pack!

Wildlife Population Models: How Can We Use Them
for Climate-Change Projections?

In Chapters 4–6, we addressed the analog question ("how can we use them…?") by adding layers progressively onto climate simulation models. Models become increasingly complex and multi-layered as we add eco-hydrology, vegetation, and then disturbance to projected climate. Can we add yet another layer, simulating animal *population dynamics* (reproduction, growth, mortality, dispersal) on the output projected by landscape disturbance models?[52]

Over the years, wildlife population models have evolved in different conceptual frameworks from the models we met in previous chapters. Both "process-based" and empirically based models abound, but their structures are not grafted easily upon the other models that we have discussed. Without diving into the details,[53] I will say here that animal models reflect, either directly or indirectly, that most animals move, in time frames from daily to annual to once in a lifetime, unlike plants. It turns out to be more tractable to use an explicitly two-step process to project animal populations in a warming climate. First, results from a landscape disturbance model are used to calculate *metrics* of the habitat quality[54] of the animal(s) of interest. Many analyses stop there, for example with comparisons of future to current quality. Those who go further use or transform the habitat metrics to be *parameters* in process-based, often very mathematical models.

The same caveats that we saw in previous chapters apply to projections of animal populations. They rely on an edifice of vegetation and disturbance models to estimate habitat quality. A further uncertainty arises from the imperfect *correlations* between habitat *per se* and population dynamics, the latter being affected by other forces such as intra-specific competition for reproductive success and intra-specific

---

[52] On top of vegetation models, on top of eco-hydrological models.

[53] It would take us far afield to cover this topic well. Briefly, process-based models of animal populations usually have highly structured mathematics; for example, sets of *differential equations* (any equation that specifies changes over time) or *matrices* (plural of "matrix", two-dimensional mathematical structures to which some, but not all, of the rules of algebra apply). These mathematical frameworks are more attenuated than the "wiring diagrams" or flowcharts that describe process-based vegetation models, for example. On the empirical side, animal models are more similar to those for vegetation, except for one crucial difference: because animals move, the idea that "absence of evidence is not evidence of absence" is invoked more explicitly. For example, even the simplest models usually have two parts. One asks "is it there?" and the other "if so, how many?" Some resources, among many, are R.M. Dorazio. 2007. On the choice of statistical models for estimating occurrence and extinction from animal surveys. *Ecology* 88:2773–2782, and (not for the casual reader) H. Caswell. 2001. *Matrix Population Models*. Sinauer Associates, Sunderland, MA.

[54] For example, a model named "HABIT@" (yes, really—see https://www.umass.edu/landeco/research/habitat/habitat.html) calculates quality at three characteristic scales corresponding to levels of biological organization: "local", "home range", and "landscape". The averages and *variances* of the three numbers obtained can be compared among time periods.

and inter-specific competition for the resources that are aggregated into metrics of habitat quality.

Models based on habitat quality are much more useful for projecting the futures of habitat *specialists* than for habitat *generalists*. For example, future losses of snow cover or old forest are bad news for the Canada lynx and the wolverine, or for the NSO, respectively. We can't make analogous statements for the grizzly bear or the gray wolf, or for the mountain lion and coyote, even though "poor habitat" or "lack of habitat" are still meaningful for these generalists. Many more *variables*, for example human acceptance or tolerance, should be considered when projecting their futures.

What Do We Expect for the Western Mountains?

To make educated estimates for members of the animal kingdom in the Western Mountains, we can ask general questions, and see how each applies in the different mountain ranges, while recognizing that one could ask all the questions about all the ranges.[55] "Who needs snow?" "Who needs forest?" Who is sensitive to elevated temperatures?" "Who is sensitive to drought?" "Who is sensitive to disturbance?" "Who needs large intact[56] landscapes?" This last question applies most to generalists with large home ranges, in or among all the mountain ranges. We have explicit estimates of vulnerability, though rough, for individual species (see Note 47), but here I present just a few examples and focus on more general concepts.

The Cascade Range and the Pacific Coast Ranges (to the Klamath and Trinity Mountains)

Who needs snow? Who is sensitive to disturbance? Who is sensitive to elevated temperatures? Three of our examples above inhabit these ranges. Loss of winter snow cover is a concern for these lower-elevation ranges, and the remnant Canada lynx and wolverine populations will surely be the worse for that. Various birds (e.g., NSO and the marbled murrelet) and mammals (e.g., pine marten and fisher) are linked to undisturbed old-growth forests, so increased wildfire (or resumed logging), especially on the western slopes, would threaten these species. Some large charismatic species such as black bear (and grizzly bears if reintroduced), mountain

---

[55] Check back with Tables 3.1 and 6.1 to see the logic for highlighting specific questions for specific ranges.

[56] Intact? This term carries a lot of baggage; debates abound over what it means, and how it relates to "natural" and other loaded words. We will get some discussion in Chapter 9, but I will leave it to others to argue about meanings. Let us be content with "relatively unmodified by humans", without any precise numbers or percentages for "relatively".

lion, and gray wolf (if allowed to re-colonize) are habitat generalists and climate *per se* is unlikely to be a threat to their persistence. Projected elevated stream temperatures are a threat to fish populations, as they are throughout the Western Mountains.

The Northern Rocky Mountains

Who needs snow? Who is sensitive to disturbance? Who is sensitive to elevated temperatures? Even though loss of glaciers is the most visible effect on snow, loss of winter snow cover at lower elevations is critical for our example the wolverine. There will be winners and losers in the disturbance game. For example, some native birds, such as black-backed woodpeckers and flammulated owls, thrive in disturbed forests. The woodpecker seeks out intensely burned forests specifically for wood-boring beetles that colonize *snags* (trees killed by fire). The flammulated owl lives in dry forests with open canopies that are maintained by frequent fire. As in the Cascade Range and elsewhere, elevated stream temperatures may be a threat to species of fish.

The Sierra Nevada

Who needs forest? Who is sensitive to disturbance? Who needs snow? Who is sensitive to elevated temperatures? Along with the rest of the Southwest, the Sierra Nevada may be a case of "forests on the brink" (see Chapters 5 and 6). *Obligate* forest species are at risk if either drought or fire kills trees and the climate is too hot for regeneration. For example, Pacific fisher populations were once connected from the Sierra Nevada to British Columbia but are now fragmented in the Western Mountains (isolated in the Northern Rocky Mountains, the Klamath-Siskiyou Mountains, and the Sierra Nevada). Fishers need large areas (home range is about 5000 acres) of continuous forest, so fires that create a patchwork of forest and burned areas are a threat. A species that probably needs snow is the very rare Sierra Nevada red fox (Fig. 7.6). Unlike the lynx and wolverine, we don't really know if the fox is snow-obligate; it may be that those are the only populations that survived the fur trade. Most sightings have been on snow. I have focused on the low elevations of the Sierra Nevada as being vulnerable, but high-elevation environments are changing too, propelled by elevation-dependent warming, as we saw in Chapter 4. As I write, there are 16 species of birds,[57] predominantly at higher elevations, that are considered vulnerable to climate change in the Sierra Nevada, although the exact reasons are unclear (i.e., whether just warming temperatures or a combination of factors). One of these is the Clark's nutcracker, which we saw earlier as an important piece of the story of whitebark pine.

---

[57] See R.B. Siegel, *et al.* 2014. Vulnerability of birds to climate change in California's Sierra Nevada. *Avian Conservation and Ecology* 9:7. https://doi.org/10.5751/ACE-00658-090107.

Fig. 7.6 Red fox in Yosemite National Park. Photo courtesy of the National Park Service

The Pacific Coast Ranges (South of the Klamath and Trinity Mountains)

Who is sensitive to drought? Who is sensitive to disturbance? Who needs forests? As in the Sierra Nevada, forests in these ranges, where they exist at all, may be close to their limits of drought tolerance, and drought-tolerant species that can re-sprout after fire may come to dominate what forest remains (see Chapter 6). At lower elevations, wildfire hazard can be extreme during Santa Ana winds. An indirect consequence of increased fire hazard in a warming climate is that land managers are using prescribed fire and *mastication* (literally grinding up shrubs and dead wood) to reduce woody fuels,[58] mainly the highly flammable chaparral shrubs. This turns out to be decimating native bird populations.[59] This problem exemplifies how with such a complex wildland-urban interface (see Chapter 6, Note 13), the effects of a warming climate in these ranges are bound up intricately with land management and policies related to development.

---

[58] Unfortunately, it turns out that this reduced hazard doesn't last very long, because flammable (often exotic) grasses invade quickly.

[59] See E.A. Newman *et al.* 2018. Chaparral bird community responses to prescribed fire and shrub removal in three management seasons. *Journal of Applied Ecology* 55:1615–1625.

The Southern and Central Rocky Mountains

Who needs snow? Who is sensitive to elevated temperatures? Who is sensitive to disturbances? With warming temperatures, winter snow will move upward and in some cases disappear altogether, leaving snow-obligate species bereft. The white-tailed ptarmigan is a master of camouflage, with a coat that is streaked brown and gray in the summer but pure white in the winter. "Standing out" in snow-free land-scapes in winter is less of a concern than heat-sensitivity; these birds often roost in the snow to cool off, especially in temperatures above 70 °F (in summer, when they are not white). Two species of weasels and several species of bats, already uncommon, are associated with high-elevation forests in the Rocky Mountains and may be vulnerable to direct and indirect[60] effects of warming climate. Not all animals that are adapted to snow are necessarily at risk. For example, the snowshoe hare, the principal prey of the Canada lynx further north, is also found in arid mountain ranges such as the Sierra Nevada, and is more reliant on vegetation (thick shrubs) than abundant snow. Perhaps surprisingly, no mammal or bird species are obvious losers from outbreaks of the MPB and its relatives.[61]

The Sky Islands and the Basin Ranges

Who is sensitive to drought? Who is sensitive to elevated temperatures? As we saw in Chapters 4–6, these mountains will likely be the hardest hit ecologically by future warming. Flightless animal species associated with the higher elevations will have no options for dispersal when their current habitat shrinks or disappears in response to warming temperatures and drought. Highly mobile species (i.e., some birds) may have some success.[62]

---

[60] For example, pathogens (for bats) or larger predators (for weasels) dispersing up in elevation.

[61] For those interested in the details, see J.S. Ivan *et al.* 2018. Mammalian responses to changed forest conditions resulting from bark beetle outbreaks in the southern Rocky Mountains. *Ecosphere* 9(8):e02369. https://doi.org/10.1002/ecs2.2369 (free online), or V. A. Saab *et al.* 2014. Ecological consequences of mountain pine beetle outbreaks for wildlife in western North American forests. *Forest Science* 60:539–559 (also free online).

[62] Particularly vulnerable are bat, lizard, and salamander species that are already endangered. The jaguar, already highly endangered north of Mexico, is more vulnerable to lethal human contact than to warming climate *per se*. The more than 50 species of birds, including 10 species of hummingbirds, may or may not be able to disperse successfully. For a gallery of Sky Island birds, see https://www.skyislandalliance.org/the-sky-islands/species-gallery/birds/.

Chapter 8
Extremes, Thresholds, Vulnerabilities

Go not to the Elves for counsel, for they will say both no and
yes.
—J.R.R. Tolkien

The cast of characters in our story is now nearly complete,[1] except for the last principal, humans, who will have the lead in Chapter 9. In each case, beginning with climate itself, the most noticeable responses to a warming climate can be the hardest to measure accurately or predict, whereas broad-scale averages, such as annual temperature, are the least discernible to us[2] but the easiest to predict. For example, global-average changes in the hydroclimate, shrinking glaciers, carbon sequestration in forests, and annual area burned or defoliated by insects can be far from obvious. In contrast, the worst flood ever, or the worst wildfire, or a vanished glacier are more newsworthy, and cause more damage, but are far less predictable.

Our key concepts of *detection* and *attribution* are important here. Detection of extreme events or big changes, such as from forest to shrubland, is usually straightforward, whereas their attribution to warming climate, in whole or part, is challenging. Nevertheless, there are two ways by which we can measure the degree to which events can be attributed. The first is easier, but less satisfying: we look at both their frequency and their severity and contrast their likelihood in the future vs. the present or at some time in the past. The statistics can be convincing, but it may be difficult or impossible to tell what climate is "doing" to increase the occurrence or magnitude of these events. The second approach takes advantage of physical mechanisms

---

[1] I acknowledge again that this coverage is not exhaustive. I have said little about micro-organisms and ecosystem biogeochemistry, for example, both of which play their parts in the responses of mountain ecosystems to a warming climate.

[2] For example, as mentioned in Chapter 3, a rise in the global average temperature by 1.5 or 2 °C, barely noticeable, if at all, from day to day, may mean huge changes in the more noticeable aspects such as heat waves, sea-level rise, and human refugees from all the disruptions caused or exacerbated by warming climate. See https://report.ipcc.ch/sr15/pdf/sr15_spm_final.pdf, published as I write.

© Springer Nature Switzerland AG 2020
D. McKenzie, *Mountains in the Greenhouse*,
https://doi.org/10.1007/978-3-030-42432-9_8

that we can understand, and claims that "These events are more likely and severe because process X, caused by climate change, is driving them." This second type of attribution is very much work in progress in climate-change science. Both types are useful for understanding the frequency and severity of extreme events and the nature, locations, and extent of *ecological thresholds*[3] that may be crossed (probably irreversibly for decades, if not centuries) in a warming climate.

Extremes

Through the early twenty-first century, the global occurrence of heat waves and droughts, storm surges and hurricanes, and some unusually severe wildfires have occasioned both research and speculation that attribute them to climate change. In the Western Mountains, the key extremes are *hydroclimatic* events (e.g., droughts and floods) and *disturbances* (e.g., wildfires and insect outbreaks). A single extreme "event" can be as short as a few hours to a few days (e.g., fires and floods) or last for a year or longer[4] (e.g., insect outbreaks and multi-year droughts).

Detection of extremes ranges from subjective judgment, more or less "we know it when we see it", to statistical methods, wherein we ask how rare, or unlikely, a specific event is. The easiest call is when an event is the worst ever by some measurable criterion. For example, a wildfire could burn the largest area ever, kill the highest percentage of trees, damage the most structures, harm the most people,[5] or create the worst smoke. What about "one of the worst ten", or "in the worst 10%", or "worse than average"? A general practice is to call events that are above the 95th *percentile* (meaning that 95% of known events have been less) in the chosen criterion[6] "extreme". So what is an extreme western wildfire in the early twenty-first century, for example? If we compare the ten largest wildfires in California since 2000 to the ten most "destructive" (structures damaged), the ten deadliest (deaths),

---

[3] The concept of "tipping points", which is used frequently and is certainly more evocative, is similar. In some cases a major ecological shift, such as from forest to non-forest (see sections on "Forest on the brink" in Chapters 5 and 6), can seem like falling off a cliff. Whether tipping or crossing a threshold, these transitions are not easily reversed. For example, in a transition back to a cooler climate, there are many possibilities for a shrubland that was once trees. A seed source may not exist, even if trees, and seedlings, could once again survive in the new climate.

[4] Some readers may have encountered the terms "press" and "pulse" for different types of disturbances. These jargon terms mean pretty much what they would seem to mean: a press "leans" on an ecosystem for an extended time, with its severity tied directly to its length, whereas a pulse is short. This distinction *per se* does not help our discussion of detection and attribution, so I will not use it.

[5] Thankfully, most wildfires in the Western Mountains don't kill people, but the smoke can be very toxic to many, especially those who are frail or have respiratory disease.

[6] There are many ways to specify the "population" of events that are the basis for the 95% calculation. An important one is the *length of record*. Do we go back in time to year 1900, or 900, or just to 1950 or even 2000?

and the ten costliest (estimated insured losses) the ranks will be different, but some of the same fires appear in every list. Each of these lists is meaningful, but they complicate the detection or identification of extremes, and therefore how we might estimate their increased occurrence in a warming climate.

Detection and identification would be most useful for our purposes if we could measure the aspects of extremes that are associated most directly[7] with warming climate. For example, the loss of mature trees or even entire forests in an extended drought reflects the combination of heat stress and water stress brought on by changing climate and hydroclimate. Similarly, annual acres burned and fire severity reflect annual climate and short-term fire weather, respectively. Area affected by a mountain pine beetle (MPB) outbreak, in contrast, reflects both climatic influences and *host vulnerability*, as we saw in Chapter 6, so it is less direct as a signal of warming climate. A better signal for the MPB would be the range of elevations affected, reflecting the insect's ability to thrive at higher elevation as the climate warms.[8]

The strongest and least equivocal *attribution* that we can make of an extreme event to climate change is to claim (or paraphrase) "This was a climate-change event." In other words, it would not have happened without global warming. Some authors in the scientific literature have come close to that, for example attributing forest die-off in the Southwest to "climate-change type drought".[9] We are on safer ground if we don't go all the way to attributing single events, particularly shorter ones such as wildfires, to climate change, but instead infer that in a warmer climate they will become more common.[10]

How much more common? When we make numerical predictions of anything, a large sample is our friend, but rareness is our nemesis. There are many more droughts, fires, heavy rains, and bark-beetle attacks than there are "global-change type" droughts (see Note 9), "megafires", rain-on-snow events with flooding, or massive bark-beetle outbreaks. In Chapter 4 we saw how we can attribute some changes of hydroclimate to changing global circulation, and in Chapter 6 how we

---

[7] The alert reader may conclude that I seem to be working backwards from attribution here. To avoid circular reasoning, we don't want to pick criteria just because the numbers of extremes, using these criteria, already seem to be increasing in our current warming climate. But we want to control for any confounding variables that could be problematic for attribution. For an obvious example, if we used reduced water use as a criterion for drought, it could be confounded with changing water regulations or new (unrelated to climate) diversions from reservoirs. Less obviously but still confounding, the number of structures damaged or the costs of wildfires as criteria are confounded by increased population in the Wildland-Urban Interface (see Chapter 6, Note 13) and the increasing operational costs of firefighting.

[8] Recall that recently the MPB has begun to attack whitebark pine, a subalpine species that lives at higher elevations than its other host lodgepole pine.

[9] Well worth the read, and not too technical. D. Breashears, *et al.* 2005. Regional vegetation die-off in response to global-change-type drought. *Proceedings of the National Academy of Sciences, USA* 102:15144–15148. In other words, the sort of drought that we would expect under global change.

[10] Yes, and therefore less extreme by definition. With more events of the same magnitude, it takes the same rareness to be extreme, whether a top-ten designation or something more quantitative like a 95th-percentile designation.

can attribute changes of fire regimes to hydroclimate. These attributions work best for averages, such as annual precipitation or annual area burned by wildfire, because the *forcings* (see Chapter 3) themselves are most predictable as averages. More atmospheric rivers, on average, will be part of a hotter world in which the atmosphere can hold more moisture. More big fires, on average, will be part of a hotter drier world, in flammability-limited ecosystems (see Chapter 6) such as most forests. But predicting exactly where and when future extreme events will occur is limited currently to time scales from hours to weeks.[11]

The limits associated with rareness are inherent in the statistical approaches that are the current paradigm in the study of climate change and extreme events. With a better physical understanding of the climate system, however, we may be able to attribute single extreme events, past or future, to processes that are definitively associated with warming climate. For example, extended weather extremes such as the 2003 European heat wave, the 2010 Pakistan flood, or the recent extended drought in California may reflect the reduced temperature difference between the tropics and mid-latitudes and the polar regions, generating some stagnation in the global circulation.[12] More insights like this will come undoubtedly as climate change continues.

Ecological Thresholds

In Chapters 1 and 4, I called the transition from water to ice, or snow, and when (annually) and where (in the Western Mountains) it happens, a nearly pure signal of climate change. Melting or freezing are examples of a *phase transition*, a distinct change between two different states. With some care, we can apply the concept to ecological states, giving us some elegant mathematical methods[13] to help us identify alternate states, such as grassland vs. woodland, forest vs. shrubland or meadow, or wetland vs. upland. With analysis, rather than just a visual (e.g., are there trees or not?), we can see these state changes more dynamically and precisely. This is important for when we confront ecological thresholds, whether crossed already or expected in the future.

I shall define an *ecological threshold* as a phase transition that is irreversible (i.e., unlike ice to water) in the short term, the time scale addressed in this book (see Chapter 1). Its being irreversible is more important than how different ecosystem

---

[11] For example, flooding caused by atmospheric rivers or large fires driven by extreme fire weather. The limits are due to the chaotic nature of weather.

[12] M.E. Mann, *et al.* 2017. Influence of anthropogenic climate change on planetary wave resonance and extreme weather events. *Scientific Reports* 7:45242 (free online). Despite the scary term in the title this paper is mostly pretty accessible.

[13] Which I will omit here. For those interested, see B. Milne, *et al.* 1996. Detection of critical densities associated with pinon-juniper woodland ecotones, *Ecology* 77:805–821. Or our paper, D. McKenzie and M.C. Kennedy. 2012. Power laws reveal phase transitions in landscape controls of fire regimes. *Nature Communications* doi: 10.1038/ncomms1731. (free online).

states appear before vs. after. For example, a forest stand may lose all of its live trees in a high-severity fire, but re-establish quickly from *serotinous* cones to its (differently appearing) former state. In contrast, the barely apparent loss of just "enough" trees from a woodland can cause it to lose the remaining trees (see citation in Note 13).

Even in a more or less stable climate,[14] ecosystems can cross thresholds, both small and large, from non-climatic causes. Repeated or sequential disturbances, whether human-caused or not, can prevent an ecosystem from returning to its pre-disturbed state. For example, two (or more) wildfires at unusually short intervals[15] in a forest whose trees have *serotinous* cones can kill the trees (first fire) and eliminate the seed source (second fire) by burning juvenile trees before they have time to produce cones. In less subtle ways, humans send ecosystems across thresholds, sometimes on larger scales than are likely to occur without human influence. For example, as I write, the Amazon rainforest[16] is being cleared and burned on a massive scale. Besides influencing the global hydroclimate, the Amazon makes its own hydroclimate from plant-water interactions on that massive scale; no vegetation means much less rain. Furthermore, essentially all the carbon and nutrients in the rainforest are stored in the canopy, and soils themselves are relatively barren. So loss of the canopy derails the local (and global) hydroclimate and the biochemistry. Warming climate may exacerbate this change, but it is not the cause.

Attribution of crossed thresholds to a warming climate raises similar issues to those with attribution of extreme events. If there will be some of each even in a stable climate, what constitutes a signal of a change *caused* by climate? I noted for extremes that changes in their frequency or severity can be analyzed statistically, with implications for detection of changes and thus projections of further changes as the climate warms. What is different for ecological thresholds is that they are not *countable* in the same way as extreme events. For example, "How many wildfires larger than 10,000 acres?" (an extreme-event question) has no easy equivalent in countable crossings of ecological thresholds, because these involve not only observation, but also *inference* (How and why is this an ecological threshold?).[17] Also in contrast to extremes, we are not handicapped by rareness or small samples *per se*,

---

[14] Recall our discussion in Chapter 3. Variation is the norm rather than the exception in climate. The range of variation, such as in global annual average temperature, can be the same in a stable climate as in one with a clear trend, such as ours at present.

[15] A possible consequence of warming climate is more frequent fires in some ecosystems. If fire is more frequent, on average, the probability of this threshold's being crossed increases. But fires happen at irregular intervals, so this threshold could be crossed without any change in the average frequency.

[16] Not part of the Western Mountains, but the clearest (and scariest, in terms of global consequences) example that I know. Clearcut logging in the Pacific Northwest is another example, smaller-scale and less irreversible over time.

[17] This may seem like an arcane distinction, but there are consequences for how we project crossed thresholds into the future. There is no natural recourse to a statistical projection like "climate change will make these extremes more likely, on average". It makes us think harder about what would constitute a meaningful projection. Read on.

because we do not depend on numbers (of droughts, fires, floods, etc.) as evidence of change.[18] This helps with the problem of small samples, but ultimately we have no easy way to predict the crossing of ecological thresholds, even if we think sometimes that "I know it when I see it".[19]

What we can do is look to some of the polarities that I have associated in Chapters 4–7 with understanding the effects of warming climate. Specifically, in each category, is there or will there be a change, probably in one direction, from one to the other?

Hydroclimate

Snow to rain, or ice to water.[20] As temperatures increase, and seasonal and perennial snow cover retreats to higher elevations (where they exist), what will happen to ecosystems or organisms in the transition zone? Seasonal patterns of streamflow, as measured in *hydrographs* (see Chapter 4), could change substantially. Disappearing snow would eliminate habitat for Canada lynx and wolverine (see Chapter 7).

Forests and Trees

Energy-limited to *water-limited*. As systems become more water-limited, species composition and forest density will change, possibly quickly and significantly. In the most water-limited zones, forests could disappear, possibly quickly, especially if there is a coincident major disturbance (see "Forests on the brink", Chapters 5 and 6).

---

[18] Certainly one can imagine a similar use of numbers to counting extreme events, such as counting the number of forest to non-forest transitions within a year, or a fixed area, or both. But this will invoke criteria for identifying a crossed threshold that are more subjective than a fire size. For example, what proportional reduction in tree cover constitutes crossing a threshold away from forest?

[19] In saying this I take issue with some recent technical papers that have been published in prestigious journals, claiming that there are statistical signals in ecological dynamics that portend tipping points or crossed thresholds. These papers are cited widely, but they suffer from "cherry-picking", choosing data in a way that supports your claims. I won't list any of them here, but for those interested in a general caution against such claims and willing to dig into some technical details, see A. Hastings and D.B. Wysham, 2010. Regime shifts in ecological systems can occur with no warning. *Ecology Letters* 13:464-472. An equally prestigious but less visible journal.

[20] Recall from Chapter 4 that there is a complication here. If precipitation increases, even with warming there could be more total snow, but it would be "concentrated" at higher elevations. Total volume in a hydrograph might increase, but the change in seasonal pattern is the potential threshold here.

Disturbance (Wildfire and Insect Outbreaks)

Flammability-limited to *fuel-limited*, and *endemic* to *epidemic*. For wildfire, the crossed threshold would be at a broader scale than the others here. Recall that flammability-limited systems always have enough fuel to sustain a fire, and require only an ignition and "fire weather" (hot, dry, windy, often all of these). Fuel-limited systems require some pulse in fine fuels (essentially kindling), usually grasses, to carry a fire through gaps in vegetation. Recall also that the flammability-fuel change often parallels that from energy limitation to water limitation. Because this threshold could be crossed over large parts of the West, we could see a major shift in landscapes that are at the most risk for wildfire,[21] with big implications for management and policy. In Chapter 6 I discussed the change in population dynamics of the mountain pine beetle (MPB), from *semi-voltine* to *uni-voltine*, as annual average temperatures increase. In conjunction with *forest succession* and changes in *host vulnerability*, areas of susceptibility to outbreaks, and their aftermath, will move to higher elevations. There is less total area up high, so vulnerable area will actually decrease.

Wildlife

Connected to *fragmented* landscapes. When an animal uses most or all of its *home range* regularly, whether it is a salamander in a small stream reach or a bear over hundreds of acres, passage within the home range needs to be feasible. Although the most obvious "fragmentors", such as dams or highways, may be human-caused and unrelated to climate change, others, such as loss of forest cover that conceals an animal from predators, or a change from year-round to seasonal streamflow that blocks a fish migration, can be just as lethal.

From these examples, and some of those in previous chapters, you can see that the causes of crossed thresholds can be synergistic. For example, a forest could become more water-limited, less flammability-limited, and more vulnerable to bark-beetle outbreaks. We might expect a major change, but what would it be? (Hint: I don't know, but bets are that the answer will not be simple). Will that new ecosystem be part of a more or less fragmented landscape,[22] and for which species?

---

[21] So here I am pushing back on the current paradigm that all wildfire hell will break loose across the West in a warming climate. Indeed there will be areas that are bound to see increased fire, as the flammability limitation weakens, but in other areas the fuel limitation will likely increase. For details, see our papers (authors McKenzie, Littell, and others) that I have cited in previous chapters.

[22] For completeness, I should note something else that goes against common wisdom. Fragmented habitat, with the same total area, is not necessarily a problem for animal species. It is only when that fragmentation is within an animal's (or a pack's or herd's) regular range of movement that it causes a problem. For a technical discussion, see L. Fahrig, 2019. Habitat fragmentation: A long and tangled tale. *Global Ecology and Biogeography* 28:33–41.

Crossed thresholds are not always associated directly with extremes. Clearly they can be, but they can also be the result of a slow steady *forcing*, which at some point becomes the proverbial final straw. Two such (aforementioned) slow forcings are rising stream temperatures that are lethal to fish and rising atmospheric temperatures that shift the reproductive frequency of the MPB. Nor does a crossed threshold that is caused by an extreme have to follow immediately, sometimes confounding our understanding of cause and effect. A good example of this is "extinction debt", which can be precipitated by an extreme or by a slow change. A case of both, sequentially, which I mentioned in Chapters 6 and 7, is *refugial* habitat for the Northern Spotted Owl on the eastern slopes of the Cascade Range. Massive clearcut logging in its original habitat on the western slopes slowly brought the owl toward an extinction threshold, but a significant number of owls were able to disperse to the eastern-slope refugia. An extreme event, i.e., exceptionally widespread wildfire across that area, could finish it off.[23]

How Can We Anticipate Change?

The Western Mountains will see big changes in the coming decades. Some may happen continuously, whereas others may be abrupt crossings of thresholds, whether or not precipitated by extreme events. Without drastic mitigation of greenhouse-gas emissions, they may accelerate by mid-century, as I noted in Chapter 1. In Chapter 9, I will ask how we can manage those big changes, including those that have begun already, but to plan for that we need to identify where, how soon, and how much we expect the changes to be. We have two distinct toolkits for this work. The first is *field-based* knowledge, in the broadest sense[24]: what we can discover from being "out there" on or above Western-Mountain landscapes. The second comprises *models*, some of which, the statistical, mathematical, and simulation models, we have met in previous chapters. In the next section we will see how *conceptual* models can tell us how to construct the other types.

---

[23] But not necessarily. We still don't know the fate of the owl, because extinction debt may not come due all at once, and those that survived fires might find less suitable refugia that still sustained them just enough. If logging had not been curtailed in time, however, extinction was near certain.

[24] More and more, this will include *remote sensing*: any manner of looking from afar ranging from instruments on low-flying aircraft, or even drones, to satellites. So "out there" can be way out there; "the field" extends into space.

Field-Based Knowledge

Change detection is the term that covers scientific work to identify where and how fast mountain ecosystems are changing.[25] *Monitoring* is any organized effort that is dedicated to ongoing change detection. On the ground, it means repeat visits to field plots to re-measure attributes of interest, anything from snow cover to tree diameters to bird nests (Fig. 8.1). With *remote sensing* (literally, any technique that isn't "right there"), it means collecting and processing data that range from images to chemical emissions.[26] Choosing how to monitor[27] involves trade-offs between measuring the proverbial "forest" vs. its "trees". The farther away an instrument for change detection, whether a human eye or some other detector, the fewer details will be captured. In exchange, it covers a larger area, and can detect changes in larger-scale structure. For example, ground-based plots will measure numbers of trees and shrubs, and how dense they are, whereas remotely sensed images will provide the spatial structure of patches of forest vs. shrubland. As I have described earlier, both are important indicators of the state of mountain ecosystems.

Science is a big part of the field-based toolkit, but not the only part. More experiential ways of knowing can bring insights about longer-term and larger-scale changes that are not evident from detailed scientific measurements. For example, a vast and diverse base of indigenous knowledge[28] has persisted and grown in Native American communities around the West for millennia. Far more than just having names for plants and animals, or knowing the timing of salmon runs or caribou migrations, indigenous knowledge includes a profound understanding of ecological processes, with all their associated complexity. Others[29] are far more literate in indigenous knowledge than I, but I would argue that such a holistic or *systems-based* approach, even if not numerically rigorous like some of our models, is a

---

[25] Not only in response to warming climate, of course. Scientific change detection is useful for measuring human-caused changes and their consequences, and for understanding natural variability arising from multiple factors.

[26] For example, satellite-based change detection compares images gathered over time for differences in *spectral signatures*, proportions of incoming radiation (light from sources on the Earth's surface) at different wavelengths. A forest has a very different spectral signature from a clearcut, or a shrub field, or a recent fire. Monitoring stations and networks, such as IMPROVE (see Chapter 6, Note 24), record emissions from fires and other origins that have been transported in the atmosphere from their sources. Repeat photography, as we saw with glaciers in Chapter 4, can be a valuable low-tech signal of long-term change and whether it is accelerating.

[27] And more generally, how to detect ecological changes. Because monitoring programs are unfortunately a low priority for agencies managing public lands in the West, much of what we know comes from shorter-term research that focuses on specific questions.

[28] Sometimes called "traditional ecological knowledge".

[29] A recent tour-de-force of this is Robin Wall Kimmerer's *Braiding Sweetgrass: Indigenous Wisdom, Scientific Knowledge, and the Teachings of Plants*. 2013, Milkweed Editions, Minneapolis. A paper in the scientific literature that analyzes the crossover between scientific knowledge and indigenous knowledge is J.O. Yazzie, *et al.* 2019. Diné kinship as a framework for conserving native tree species in climate change. *Ecological Applications* 29:1331–1343.

Fig. 8.1 Monitoring can be exciting. Sampling the lake in Loch Vale (Fig. 2.7) in winter. Photo by John Hammond

necessary complement to the scientific method[30] for understanding something so complex as the effects of climate change on mountain ecosystems.

Anecdotal knowledge (i.e., not having been through a rigorous analysis) can also be valuable if it is not just accepted as truth uncritically. Observations like "The birds aren't singing in the morning any more", or "This stream has never run dry in August before" are clues to the consequences of a warming climate that may have escaped more systematic change detection, if for no other reason than that "detectors" can't be everywhere at once. Observant residents of the Western Mountains and those (like your author) who have spent decades traveling and wandering there are an increasingly valued resource for gaining a comprehensive picture of the ongoing effects of warming climate.

---

[30]Yes, there are certainly systems approaches in science too, along with "mechanistic" or "reductionist" approaches that try to isolate causes and effects. I encourage readers with some literacy in both areas to imagine the boxes and arrows of a formal scientific model of an ecosystem vs. those in a model based on indigenous knowledge. Especially in the "arrows", can we claim that there is less uncertainty and ambiguity in the scientific version? Think of the models we have discussed so far.

Much More About Models

We have encountered many model types in previous chapters, and some discussion about their value, but now I want to take one step back and frame the idea of a "model" more generally. This section (up to "Vulnerable landscapes") will be slower going than the rest of the book.[31] We can then revisit the specific ranges in our Western Mountains one more time, how they may be confronted with abrupt changes such as extreme events or crossed thresholds, and how well (and for how long) we can anticipate future change with models. Just before that, we will note the intrinsic limitations of future projections, some of which I introduced in Chapter 4.

All disciplines, not just science, build models of reality. This includes fiction, history, humanistic scholarship in general, and psychoanalytic theory, to name just a few examples. In all of these,[32] both aesthetics and methodical analysis of data contribute to models' success and acceptance. In scientific models, more than the others, correspondence with data is not optional, and aesthetics take second place.[33] On our topic—the future of the Western Mountains—we are obviously limited further in that our models can be "confronted"[34] only with current or past data.

Conceptual models create a framework for thinking about how processes that we understand interact. They can have few numbers and abundant logic, or the reverse, or much of both, but their purpose is to codify our thinking about a problem to be solved, but not necessarily to produce new numbers, or answers, or future projections. For example, a *stress complex* is a conceptual model of how multiple factors interact to make landscapes vulnerable to ecological change.[35] As shown in the figure for the Sierra Nevada (Fig. 8.1), the *forcing* from global warming sets in motion a variety of ecological changes with one or more consequences that are generally unfavorable. We can see from the example that simple cause-and-effect models can miss much of the complexity, and may thereby miss the outcome. Global warming interacts with human-caused fire exclusion and ozone pollution to exacerbate fire

---

[31] So feel free to leap ahead, and return to this later, or not.

[32] You might not agree about fiction. Perhaps not all fiction, but think of historical fiction, such as novels set in the Roman Empire. Some data are necessary.

[33] Notwithstanding some recent excursions in theoretical physics, in which the "naturalness" of theories has re-emerged as a basis for evaluation. For contrasting views that are very accessible to the lay reader, see F. Wilczek, 2015. *A Beautiful Question: Finding Nature's Deep Design*. Penguin Books, New York, and S. Hossenfelder, 2018. *Lost in Math: How Beauty Leads Physics Astray*. Basic Books. New York.

[34] This expression comes directly from an excellent book, which is however not for the casual reader. R. Hilborn and M. Mangel. 1997. *The Ecological Detective: Confronting Models with Data*. Princeton University Press.

[35] Full disclosure: we published this idea in 2009 in an obscure book that is now out of print. Therein we described models for pinyon-juniper landscapes of the Southwest, spruce forests of Interior Alaska, lodgepole pine forests of the Rocky Mountains, and the example in the figure, the Sierra Nevada. The chapter is pages 319–337 in S.V. Krupa (ed.), Developments in Environmental Science, Vol. 8, Wild Land Fires and Air Pollution, A. Bytnerowicz *et al.* (eds.). Amsterdam, The Netherlands: Elsevier Science, Ltd.

severity and tree mortality, with a changed or even eliminated forest as the consequence (Fig. 8.2).

   Scenario planning is a type of conceptual model that emphasizes the widest
plausible range of outcomes. It is the purest form of "what if?", which we met as the
question asked by all future projections (rather than "what will be?"). Two of the
three socioeconomic inputs[36] to climate models that I discussed in Chapter 3 are
scenarios. As with other conceptual models, scenario planning focuses on getting
the logic right. It trades off (in most cases) the detailed outcomes of one scenario for
the broad coverage of different scenarios, and by doing so embraces uncertainty
instead of trying to minimize it. Scenario planning has found particular favor with

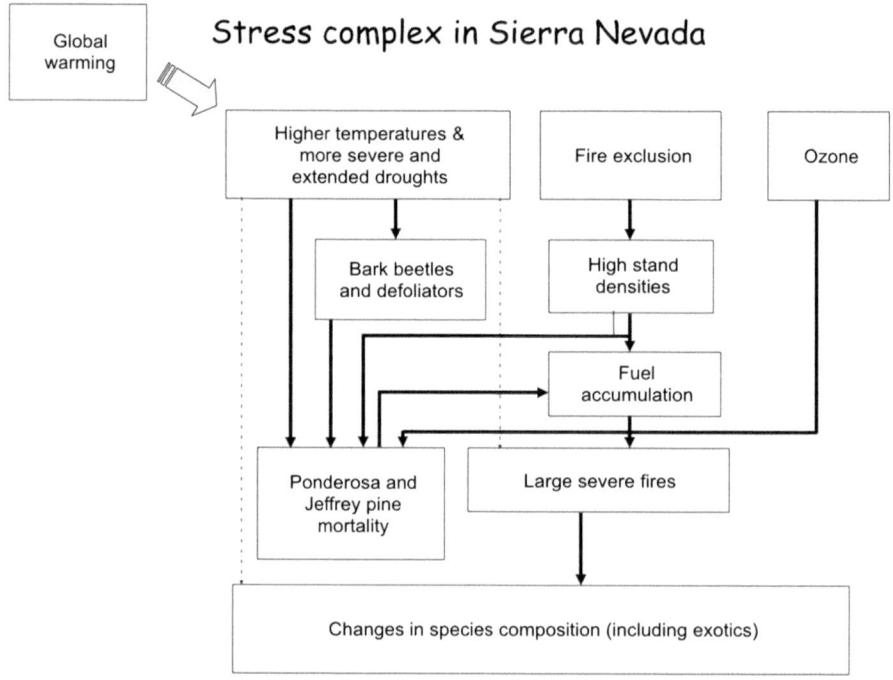

Fig. 8.2 Stress complex in Sierra Nevada and southern Californian mixed-conifer forests. The
effects of disturbance regimes (insects and fire) and fire exclusion are exacerbated by global warming. Stand-replacing fires and drought-induced mortality both contribute to species changes and
exotic invasions

---

[36] The SRES (Special Report on Emissions Scenarios), the first group used with climate projections, and the SSPs (Shared Socioeconomic Pathways), the most recent group, are of the "what if?"
type. The RCPs (Representative Concentration Pathways) are "what had to be" to produce various
levels of *radiative forcing* by 2100. This *inverse-modeling* aspect of the RCPs has made them
harder to understand for many. All of these scenarios are from the IPCC; they are still trying to get
it perfect. See discussion in Chapter 3.

those working on adaptation to climate change: "what do I do if?" More on that in Chapter 9.

Abstraction and Complexity

> *As simple as possible, but no simpler*
> —attributed to Albert Einstein

> *The fancier the plumbing, the easier it is to stop up the drain*
> —Scotty (chief engineer on the starship Enterprise)

If we want to build the best possible model to project future (mountain ecosystem) changes, it may seem intuitive at first to include everything we know about how the (eco)system works. We don't want to miss anything that may throw off our estimates. This inclusiveness will take different but related forms in *inverse modeling* versus *forward modeling*.[37] In inverse modeling, we are usually seeking to identify *variables* that are predictive of the *responses* that we observe and want to project into the future. In forward modeling, we want to specify relationships, usually causal, between variables that we have already chosen as important. Both inverse and forward modeling are subject to tradeoffs along two dimensions, or "axes" (Fig. 8.3): *abstraction* (abstract vs. concrete) and *complexity* (complex vs. simple).

The most inclusive inverse model would specify every variable that might have some effect on the response. For example, suppose we want to project how a tree species will respond to a warming climate. How many variables are important? We certainly expect that temperature itself will make a difference. What about seasonal vs. annual average temperature? What about maximum vs. minimum temperature, either their peaks or their averages? Related but different variables, such as drought, or solar radiation, can be parsed in similar ways. Other variables that depend on temperature, such as snow cover or depth, may have a more direct influence than temperature itself. There are many ways to "pile on" variables until we have included everything, and then start to prune back those that do not have enough of an influence to "matter".[38] We may also have to consider how the remaining variables interact in their effects. For example, does temperature matter more when it's wet, or dry? (Conversely, does rainfall matter more when it's hot, or cold?). Our best model

---

[37] Refer to Chapter 4, the section on "ecohydrological" models, for explanations of these terms.

[38] There is a vast statistical literature on this, some of it very mathematical. To take just a dip here, probably the most obvious "excess" of *predictors* occurs when two or more variables are surrogates for the same underlying influence. For example, annual average temperature, summer average temperature, and average daily maximum temperature may (or may not) be measuring more or less the same effect. If they are strongly correlated, they are more likely to be redundant. A heuristic process that modelers sometimes use is to identify *proxy sets* (groups of variables that probably measure more or less the same effect) and include only one element of each set in any model.

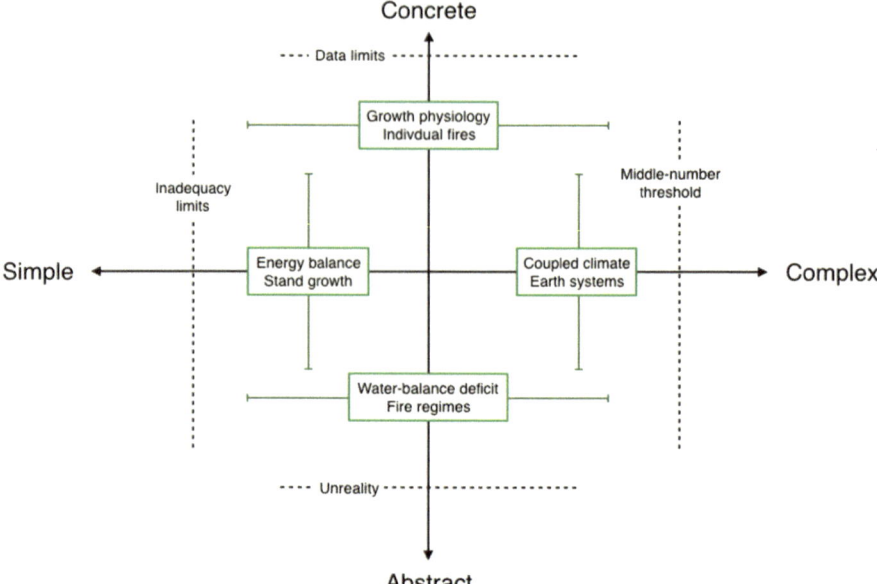

Fig. 8.3 Abstraction and complexity are independent attributes of models, in that one does not necessarily entail the other. Where your model fits in this two-dimensional space depends on the questions you are asking (e.g., "What if?"), how fast your computers are, and how good your data are. Our Einstein quote is good guidance for position on the Simple/Complex axis. I have no equally pithy formula for abstraction, but see discussion in the text for some examples of choosing a level of abstraction. On each axis, there are hard limits outside of which models will not function. Model types in boxes are fixed on one axis, but vary on the other. For example, per the text, models using water-balance deficit or estimating fire regimes are more abstract than those that simulate physiology or individual fires, but each type can be simple or complex. An analogous contrast obtains on the complexity axis

will be at the optimal[39] place on the complexity axis (Fig. 8.2). Typically this will not be "as complex as we can manage with our resources, e.g., computational, logical, or financial".[40]

In the figure we have two axes, which are *orthogonal* (perpendicular graphically, independent mathematically, i.e., you can't know from a model's complexity how

---

[39] And where is that, exactly? You should not be too surprised to hear that there is no one way to identify such a position. And that there can be many different ways to identify such a position, depending on one's specific goals for a model. Using our example of a tree species, we may care about its regional distribution in a mountain range versus whether it will thrive in a particular watershed.

[40] Rather, we follow Einstein, quoted above. At a certain point, long before we are close to mimicking reality, more complexity will generate more error and uncertainty, not less. With inverse models, this is known as "overfitting". Typically, our model may correspond more closely to observations than would its simpler counterpart, but be less robust when used for projections into the future.

abstract it is, and *vice versa*). On the Abstract-Concrete axis, we choose the level of abstraction based on both the structure of our data and our conceptual model if we have one. For example, in Chapter 6, I wrote of *fire regimes* vs. individual fires. A model that examines fire regimes will be more abstract than one for individual fires, just as the variables themselves differ in their levels of abstraction. Perhaps the clearest general example of abstraction in climate-related modeling is the use of *indices* (metrics calculated from two or more variables) that represent the aggregate effect from multiple (often *limiting*) factors. For example, the *water-balance deficit*, which we met in Chapter 4, and various drought indices,[41] have proven especially useful for aggregating the effects of individual variables.

Forward models are subject to the same considerations. The most inclusive forward model would specify every process that we consider important, in the framework of the scientific discipline in which we are working.[42] These processes are climate, hydroclimate, plant and animal ecology, and disturbance ecology. Choosing the right levels of complexity and abstraction will depend, as with inverse models, on our objectives, our conceptual model if there is one, and the structure of our data. These very general criteria cover what often becomes many hours of brainstorming, a lot of trial and error, and acceptance of uncertainty and limitations, not all of which can be overcome by better data and more computing power. Drawing on our earlier discussions, let me reiterate two concepts that are central to model projections for mountain ecosystems: interactions and limiting factors.

Interactions

Somewhere between trying to capture each individual process or event[43] and conceding that "everything is connected to everything else" is a sweet spot that we need to find. Some interactions are critical, needing to be specified, whereas others may be only incidental, and can be left to emerge from the outcomes of individual processes. Two critical interactions that we have seen earlier are (1) between drought and temperature and (2) between wildfire and bark-beetle outbreaks. Focusing on the second of these, in a landscape disturbance model (see Chapter 6) we need to constrain our outcomes based on what we know about the interaction *per se* rather

---

[41] You would not be wrong say that water-balance deficit is also a kind of drought index, one that specifically considers plant-water relationships.

[42] So that we are not simulating the entire universe. For example, we know that quantum-mechanical processes are the basis of everything, and that ecological processes do not violate the laws of physics, but even the most complex simulation model cannot be responsible for every influence at a more fundamental level. This is part of choosing the right level of *abstraction*.

[43] My personal view is that this is a key limitation of "agent-based" models, in which all outcomes ensue from certain "actors" (agents), and interactions are *emergent* from individual actions. Sometimes it is more effective, and parsimonious, to specify the interactions rather than the "actions".

than about the processes separately.[44] We thereby increase the level of *abstraction* (one higher-level "process" subsuming two that are more concrete) while managing the *complexity* (essentially modeling one process instead of two). For example, to implement this in a landscape disturbance model, we might specify the interactions between wildfire and insects by placing some constraints on their cumulative effects rather than "letting" each go its separate way. Whenever you simulate a wildfire, don't ignore what the insects are doing, and *vice versa*. Careful, though. Those constraints have to allow for extremes to *emerge* from the interactions.

Limiting Factors

If we constrain our outcomes to be consistent with what we know about limiting factors, we can sometimes save a lot of effort in making our model produce realistic outcomes.[45] I had a personal experience with a landscape disturbance model that failed to do this. The model was unmistakably complex, with detailed simulation of the physiological responses of tree species to climate, but no constraint from limiting factors. In a watershed in the Cascade Range with rugged topography, the model swapped the landscape positions of two species associated with water-limited and energy-limited environments, respectively, because it had no way of "knowing" these species' range limits. Instead, it relied on more concrete (i.e., less abstract) aspects of physiology to simulate the species' survival and growth in response to climate, but got it wrong. Here there was excessive *complexity* (in the physiological details) and not enough *abstraction* (in understanding the ecological context).[46]

---

[44] Meaning what exactly, in practice? We could let fires ignite and burn (virtually, in the model) and beetles reproduce (in galleries in the bark) and kill trees in one *time step*, without considering whether fires made trees more vulnerable or beetles made trees more flammable. This ignores the interaction, and you can see why our model might go off the rails fairly rapidly thereby.

[45] Not to be confused necessarily with expected outcomes. Sometimes a realistic outcome can be a surprise, and be very educational. Furthermore, an unrealistic outcome can also be educational. One thing we need to avoid is to forbid our models from being "wrong", especially in their early stages. Wrong answers can inform both our methods and our questions.

[46] For those who are wondering, there may have been other problems too. One may involve *scale*. Some aspects of microclimate may have been given more importance than they should, and aspects of macroclimate and limitations of dispersal given less. The lower-elevation (water-limited) species might have survived in warm pockets higher up, if it could have dispersed there, and the higher-elevation (energy-limited) species in cold-air drainages lower down, if it could have dispersed there. Modeling is difficult.

The Problems that Won't Go Away[47]

Sensitivity to initial conditions, or the "chaos" problem. One of the places that we understand this problem best is the Earth's climate. At the broadest scale, global annual average temperature, *chaotic dynamics* (physical processes that are subject to this sensitivity) are less of a problem than they are at finer scales, because things more or less "cancel out". Once we bring in atmospheric circulation at finer scales, with its effects on regional and local weather (and average weather, i.e., climate), irreducible uncertainties appear, and these can propagate into very different projections of regional climate. As we saw in Chapter 3, this is one reason that we use *ensembles* of regional and global climate models. The uncertainties associated with this problem are unlikely to disappear, however, or even decrease very much, even with much better computers[48] or understanding of the physics of climate.

Things change at different scales; part one, or the "middle-number" problem. I wrote in Chapter 4 that often what we care about is at a different (usually broader) scale than what we have a good model for. If that scale is broad enough, such as the whole Earth, taking an average can often tell us most of what we want to know, but there is a middle ground that is too complex to assess by adding up events at smaller scales, but too variable and unpredictable to use averages (of anything: climate, tree growth, animal populations, etc.). For example, we can estimate the extent and intensity of one wildfire (fine scale), and predict the annual area burned by wildfire at a regional scale reasonably well (broad scale), but "in between" predictions are difficult to impossible.[49]

Things change at different scales; part two, or the "coarse-graining" problem. One of the ways that we try to make the middle scales tractable for modeling and analysis is to add things up and calculate some believable average with the least uncertainty or error. Unfortunately, no roadmap for this process exists for the real world of the Western Mountains. As complexity (e.g., of a mountain landscape) increases, we are forced not only to aggregate individual elements and processes, but also (usually) to increase the level of abstraction with which we represent them. For example, recall my anecdote (above) about the landscape disturbance model that failed at the middle scale of a watershed, partly because the model was too concrete (not invoking the broader idea of limiting factors) and partly because of too much variability, making averages inadequate.

---

[47] For a detailed discussion of the hard problems, see the reference in Chapter 6, Note 85.

[48] So why won't better computers help? Aren't we getting longer-term weather forecasts than we used to with older computers? Yes, they help with a few extra decimal places, but with any more than a few we run into problems with "computational complexity": some problems become essentially uncomputable as the number of elements (or decimal places) increases.

[49] This "in between" refers to both spatial extent and level of *abstraction*. For extent, it is the middle ground between the extent of one fire and the size of ecoregions, or *ecosections* (see Fig. 6.1). For abstraction, it is between the most concrete, a single fire, and the most abstract, the broadest element of a fire regime, area burned. Notably, fire severity (for a fire regime, not its estimation for a single fire) and the spatial patterns of fires present a middle-number problem for prediction.

The stationarity problem. When climate itself is not stationary, not only do variables (e.g., annual temperature) have different values, but so do their effects on each other. In particular, controls on processes such as *limiting factors* change. For example, we saw in Chapter 5 how an ecosystem could change (and many probably will) from primarily energy-limited to water-limited, thereby also changing species' chief limiting factors. The stationarity problem increases unremittingly as climate continues to warm.

As we look further into the future, all of these problems, but especially the chaos and stationarity problems, will increase, along with other issues associated with more distant forecasts, such as more unforeseen extreme events and crossed thresholds. Even our most confident predictions, in this book or elsewhere, have limited lifespans. For the Western Mountains, I consider my "predictions" beyond about 2060 to be little better than random, as I said in Chapter 1, with those for some of the most vulnerable landscapes expiring sooner. With some help from mitigation and adaptation, however (see Chapter 9), some of them might be wrong in the right way.

Vulnerable Landscapes: Where in the Western Mountains?

Let us take one more pass through our mountain ranges, on whose futures the previous five chapters have ended. Which of them are the most vulnerable[50] to a warming climate, and in what ways? Some general types of changes that we might expect are:

Warming and Drying

Projected changes in circulation and the narrowing of temperature differences between the polar regions and the tropics will cause some western regions, and their mountain ranges, to warm up and dry up more than others (see Chapters 3 and 4).

Reduced or No Capacitance

This is the loss of buffers against rapid change, such as glaciers, snowpack, or forest cover. A less obvious capacitor may be biodiversity, whose loss, especially of keystone species (see Chapter 7), could disrupt food webs, *phenology* (the timing of seasonal changes in plants), and pollinators that rely on plants.

---

[50] Referring broadly to things getting "worse", and so I focus on elements of mountain landscapes that may be lost, e.g., snow or glaciers, species, both plant and animal, forests (or meadows), streams, clean air. In Chapter 9 we will look at *ecosystem services* (things that humans use) that may be lost. This is not to claim that there will be no gains of ecosystem elements that are valued, but as I suggested in Chapter 1, I believe that the average change will be toward impoverishment.

Ecological Disturbance

Disturbance *regimes* will change, in direct response to warming and drying and indirectly with changes in vegetation. Fire is likely to increase in severity and extent, as will *biotic* disturbances caused by "winners" such as the MPB, but not everywhere.

Crossed Thresholds

Two factors are generally important, (1) How much change is there to the current state, whether climate itself or its effects on ecosystems and disturbances? (2) How close is the threshold to the current state? Is there any "breathing space" for what is at risk?

Changes to Iconic Landscapes

National Parks, monuments, wilderness, and other protected areas are a patchwork across the West. Many are small, isolated, and at higher elevations than other public lands. Elevation-dependent warming, loss of *connectivity*, and extirpations of rare or *endemic* (restricted to a small geographic range) species may all be problematic.

In general, vulnerability in the northern ranges (Cascade Range, Pacific Coast Ranges North, Northern Rocky Mountains) may be greatest at the higher elevations, whereas for the southern ranges (Sierra Nevada, Central and Southern Rocky Mountains, Pacific Coast Ranges South, Sky Islands and Basin Ranges) it may be at lower elevations. Very broadly, this corresponds to the contrast between energy-limited (mostly in the North) and water-limited (mostly in the South) ecosystems.[51] Overall, vulnerability will be greater, and will come from more directions, in the southern than in the northern ranges (Table 8.1).

The Cascade Range and the Pacific Coast Ranges (to the Klamath and Trinity Mountains)

Partly buffered by the maritime climate, these ranges are probably less vulnerable than those more inland. Perennial and seasonal snow may retreat to higher elevations without total snowpack decreasing (if winters become wetter). Their drier eastern slopes will be the most clearly vulnerable to increased fire, as they are still mostly in the flammability-limited domain, except for the lowest elevations. On

---

[51] Put another way, in currently energy-limited systems, agents of change get MORE "energy". Wildfires have more fuel from denser vegetation; bark beetles have more metabolic energy to reproduce faster. In currently water-limited systems, processes that need water have LESS water. Plants and animals are stressed and many are replaced (or just lost).

Table 8.1 Comparison of ranges in the Western Mountains with respect to relative rates of change in response to warming climate, (this author's) relative confidence in landscape disturbance models, and areas expected to be most vulnerable

Range	Relative rate of change (compared to average for Western Mountains)	Confidence in model projections	Most affected areas
Cascade Range and Pacific Coast Ranges (N)	Slower	Worse than average	Eastern slopes
Northern Rocky Mountains	Close to average	Better than average	High elevations
Sierra Nevada	Faster	Better than average	Low elevations of western slopes
Pacific Coast Ranges (S)	Close to average	Worse than average	Lower treelines
Southern and Central Rocky Mountains	Close to average	Average	Low and middle elevations
Sky Islands and Basin Ranges	Much faster	Better than average	Low and middle elevations

those landscapes, however, we may see a threshold crossed from forest or woodland to non-forest. Habitat for iconic wildlife, such as Canada lynx and wolverine (snow), and northern spotted owl (mature forest), already marginal, may be lost.

These ranges include four big national parks: Olympic, North Cascades, Mt. Rainier, and Crater Lake, along with numerous wilderness areas that are mostly at higher elevations. Crater Lake, as the smallest and most southerly of the parks, is the most vulnerable to major change in the coming decades.[52] Its dense forests will be subject to increased ecological disturbance. In contrast, alpine meadows, with their renowned fields of flowers, may be invaded by trees, something we have seen already at Mt. Rainier and elsewhere in the range.

The Northern Rocky Mountains

Our other northern range has a continental climate, and "business-as-usual"[53] projections suggest as much as 4–5 °F average increase in annual temperatures by 2100. The two biggest changes (long before 2100) will be loss of snow and ice and increased ecological disturbance. Our best models of future fire regimes[54] for the Western Mountains are for this region, and they suggest substantial increases in annual burned area in a warming climate, with the fire regimes' being

---

[52] Remember the caveats above and in Chapter 1 about the short period of applicability of projections.

[53] That is, the IPCC RCP 8.5, which specifies the climate forcing from CO_2 and other greenhouse gases as 8.5 watts/square meter by 2100. See Chapter 3.

[54] See Chapter 6, Note 17.

flammability-limited. The brunt of mortality from MPB will shift upward in eleva-tion with temperature-related increases in their reproductive cycle. Our best esti-mates are that elevations of 6000–10,000 ft. will be hit the hardest, i.e., the highest-elevation forests in this range. Loss of the iconic whitebark pine, a key resource for grizzly bears, is possible, but as we saw in Chapter 6 (see "Fire-induced changes in treeline"), the fire ecology at or near treelines is complex. Interactions among fire, MPB, and the blister rust that kills whitebark pine may be antagonistic (as opposed to *synergistic*), such that their individual lethal effects may be muted.

Two big national parks, Glacier and Yellowstone, may be largely transformed. Most projections suggest that by the 2040s or 2050s, there may be only one glacier left[55] in Glacier National Park. Wildfires may change the park's vegetation to much younger stands (i.e., more frequently burned at high severity). Due to its position "downwind" from fires in the Pacific Northwest, Glacier National Park may have consistently worse air quality (due to smoke) in the future. Some projections sug-gest that wildfires in Yellowstone may burn often enough that the vegetation will change from forest to non-forest.[56] These projections are, however, from the land-scape disturbance models (of the *inverse* type) that we saw in Chapter 6, with all their limitations.

The Sierra Nevada

The greatest environmental stresses in a warmer and drier climate will be at lower elevations, where disturbances are a major driver of changes in vegetation, and water from higher elevations is already in short supply. Even without more frequent fire, low-elevation forests will be stressed by drought, with seedlings the most vul-nerable stage of a tree's life. When severe fires do occur, killing many trees, seed-lings may be unable to survive, converting forests to shrublands. Reduced streamflow (i.e., loss of *capacitance*) in late summer, the most vulnerable season, because of earlier snowmelt, will make things worse.

This crossed threshold, from forest to non-forest, may have an opposite at higher elevations, with trees invading meadows. Some meadows are large, able to persist until now from a combination of (rare) flat topography and poor soil drainage that is not hospitable to trees in an already energy-limited system whose principal limit-ing factor is weakened in a warming climate.

Air quality is already a significant concern in the Sierra Nevada. A combination of agricultural haze from the extensive farmed Central Valley of California, directly upwind, with smoke from fires in the Coast Ranges, brings many days of impaired visibility throughout the mountains, and even in the Owens Valley to the east (see Fig. 2.14). A longer or more severe fire season in the Coast Ranges could make air

---

[55] The Piegan Glacier, whose lower terminus is at 8200 ft. elevation.

[56] See Notes 73 and 88 in Chapter 6. It takes ~30 years for the *serotinous* cones of lodgepole pine to mature; if burned earlier a tree cannot reproduce.

pollution nearly intolerable for park visitors, and with poorly known effects on plant and animal life.

Two national parks, Yosemite and Sequoia, and several large wilderness areas, cover most of the core of the Sierra Nevada. The most visited areas of the parks are at their lower elevations: 4000–6000 ft. in Yosemite and 5000–7000 ft. in Sequoia, and they include three groves of giant sequoias.[57] These groves are in isolated pockets of cooler air in very water-limited landscapes, so changes in regional circulation that reduce or eliminate the cold-air drainage could be lethal. Even now, giant sequoias in Sequoia National Park that are in marginal areas are losing foliage because of drought, although as yet there has been little mortality (Fig. 8.4). But the buffer is not large.

The Pacific Coast Ranges (South of the Klamath and Trinity Mountains)

The main vulnerabilities here are those already described in Chapter 6. Even seasonal snow may all but disappear, and the threshold between forest and non-forest is also close to being crossed. Wildfires are generally high-intensity and often very destructive; burning highly flammable chaparral; this is likely only to get worse. More than any other of our ranges, the future of these mountains' ecosystems will be tied closely to human demographics. The Wildland-Urban Interface (see Note 13, Chapter 6) is extensive and continues to grow with new housing developments, bringing challenges not only for firefighting but also for protecting ecosystem resources such as connectivity for wildlife and unpolluted water in streams.

The protected areas (two National Monuments and one National Park) are three small ones, none of which is truly in the Coast Ranges: Channel Islands (off the coast of southern California), Joshua Tree (4000 ft. elevation in the Mojave Desert, but near the ranges' two highest peaks), and Pinnacles National Park (in west-central California). Of these ecosystem fragments, the most vulnerable is Joshua Tree, named for its giant yucca trees. These trees are adapted to desert climate, but not to fire, which may come from invasion of flammable exotic grasses in this *fuel-limited* system.

The Southern and Central Rocky Mountains

Like the Sierra Nevada, these ranges are vulnerable at the lower elevations, but also at middle elevations at which there is substantial continuous forest. As in the Northern Rocky Mountains, ecological disturbance is likely to be the principal agent of change. Colorado's 14 glaciers are all above 10,800 ft., so are less vulnerable

---

[57] Specifically, The Tuolumne and Mariposa groves in Yosemite and the Giant Forest in Sequoia.

Fig. 8.4 Rare but unprecedented mortality of giant sequoia trees in Sequoia National Park. Photo by Nate Stephenson

than those at the (much) lower elevations of Glacier National Park, even being far-ther south. MPB outbreaks, and those of other bark beetles and defoliators, have already been extensive and are likely to continue, albeit at higher elevations where the climate has heretofore been too cold to support outbreaks.

Like much of the eastern slope of the Cascade Range, low-elevation forests[58] are still flammability-limited but close to a transition to fuel-limited. Like the low-elevation forests of the Sierra Nevada, they are close to a threshold of water limitation such that as the climate warms, seedlings may not be able to survive after a wildfire kills the overstory. A crossed threshold from forest to non-forest is also likely here.

There are three national monuments[59] in these ranges, and various wilderness areas, but only one national park: Rocky Mountain, which holds all of the range's glaciers. In large contrast to the Sierra Nevada, whose highest peaks are part of a single cordillera, the Rocky Mountains (all of southern, central, and northern) are many separate groups of mountains,[60] several with peaks over 14,000 ft., and each with distinct vulnerabilities. For example, the ranges' largest continuous wilderness outside of Rocky Mountain National Park is in the San Juan Mountains, in southwestern Colorado. The continuous forest is susceptible to massive outbreaks of spruce bark beetle, and also to high-severity wildfire. Although fire-beetle interactions are complex, and not always additive, as we saw in Chapter 6, there is potential for large areas of tree mortality with the "right" timing of beetle-killed trees and fire ignition (when foliage is dead but not fallen off branches).

The Sky Islands and the Basin Ranges

One more time, I will suggest that these ranges may be the hardest hit by a warming climate. As I said in earlier chapters, drought and warm temperatures, invasive flammable grasses, insects that kill drought-stressed trees, and isolated high-elevation habitats all play a part. If there is good news, it is that some[61] of the iconic landscapes of these ranges may be more resilient than some of their counterparts in the other ranges. Two examples follow.

By default here, the Basin Ranges include the national parks of the canyon country of southern Utah. Bryce Canyon is a microcosm therein (total just over 35,000 acres, the size of one large wildfire) without any major vulnerability that we have associated with warming climate: topography too complex for wildfire to spread

---

[58] Recall from Chapter 2 that "low" is very different here from the Cascade Range. For example, lower treeline in the Front Range of Colorado is above 7000 ft., whereas it is closer to 1000 ft. in the Cascade Range (on the east side, with no lower treeline at all on the west side).

[59] Mesa Verde, the Black Canyon of the Gunnison, and the Great Sand Dunes.

[60] Refer back to Chapter 2, but in Colorado the four most prominent are the Front Range, running N-S on the eastern side, the Sangre de Cristo, the "College" Range (whose peaks have Ivy-League names), and the San Juan Mountains.

[61] But certainly not all. For example, Sky Islands that are smaller and lower than Great Basin National Park highlighted in the text may see their iconic habitats disappear from having nowhere "up" to retreat.

Fig. 8.5 Bryce Canyon National Park, a non-loser? Photo by the author

easily, in a very fuel-limited system, and plant species resistant to drought and insects (Fig. 8.5).

Great Basin National Park has Nevada's highest peak (Wheeler Peak, 13,065 ft.), and a remarkable elevation-based sequence of ecosystems from desert grassland to alpine rock and ice (Fig. 2.12). Although *ecotones* will move upward gradually with warming climate, there is still room for everything in the predictable future, as there is in other of the highest Sky Islands. It is also so far off the beaten track[62] that it is not likely to suffer from overuse in the manner of other parks.

---

[62] Just ask Google how long it will take to drive there from almost anywhere else.

Chapter 9
Mountains and People in a Warming World

Everywhere is downstream.

—*Taoist proverb*

Where do we fit into this story? Everywhere of course; little of the Western Mountains has seen no human footprint, and much has been altered significantly and permanently. But it is not so altered that we cannot see the signals of a warming climate through the filters of human influence. Chapters 4–8 tell how those signals came to be, how they are observed, and how they may affect the future of mountain landscapes. As I said in Chapter 1, enough of Western-Mountain ecosystems is intact that we can still watch a "natural experiment" unfold in their response to warming climate.

Mountain environments have presented extra challenges to subsistence for millenia, and been subject to resource extraction for centuries,[1] whereas only in the last two centuries have opportunities for recreation been exploited. In modern societies, all of these interactions have evolved, such that mountains are now the "upstream" element of complex webs of life-support systems, economic activity, and recreation.

Ecosystem Services

Ecosystem services comprise everything that people get from ecosystems. In the Western Mountains, these may be wide-ranging: water quality and quantity, wood products, minerals (mined), forage for livestock, air quality and visibility, soil preservation, and carbon sequestration. They also include cultural resources, for both Native Americans and later human arrivals to the West. Damage to these is largely

---

[1] For example, many historical navies depended on extraction of timber, often as unsustainable as anything we have seen across the U.S. or elsewhere globally in the twentieth-century.

© Springer Nature Switzerland AG 2020
D. McKenzie, *Mountains in the Greenhouse*,
https://doi.org/10.1007/978-3-030-42432-9_9

irreversible. We have covered the basics of water (Chapter 4), wood products and carbon (Chapter 5), and air quality (Chapter 6),[2] so I can summarize their prospects as ongoing ecosystem services in a warming climate.

Water has three aspects, for our purposes: quantity, quality, and timing. For some of the Western Mountains, the northern part, the effects of warming climate on quantity may be minimal. In the arid Southwest, with projected droughts on top of warmer temperatures, there will be real scarcity, for both vegetation and ecosystem services. Seasonal peaks of water in streams and rivers will be earlier everywhere, challenging the massive and complex water infrastructure across the West as the climate warms. For example, even in the relatively "watery" region of the Pacific Northwest, there is concern that its major waterway, the Columbia River, may be inadequate at times of peak demand.[3] For more arid regions, problems will be more acute.[4] Overall, even in the face of improved technologies and sophisticated scenario planning, the fact remains that storage is imperfect. That means that the more time passes between water's arrival at storage (e.g., dams) and its use, the greater the loss. Worse *timing* necessitates more *quantity*.

Water quality depends on both quantity and timing, along with what is upstream: other water, the land, and the atmosphere.[5] In general, too much or too little, too quickly, will reduce water quality. This is true both upstream and downstream of storage. For example, disturbances such as fires, mass movements, and floods can load waterways with more water, or debris,[6] or both, than they can handle. Streams can dry up, eliminating habitat for amphibians and other animals. Below storage, demand can exceed supply, drying up waterways as big as the Colorado River and reducing the quality of what remains from salinization.

Wood products and carbon sequestration come from the same source (trees).[7] Like the aspects of water, these services interact, with specific but complicated

---

[2] I leave the discussion of the future of mining and livestock grazing to others, even though they are indeed major disturbances across much of the West. Climate change is unlikely to affect them directly.

[3] For a look at the complexity and diversity of issues around water and infrastructure in the Pacific Northwest, see work from the Climate Impacts Group, University of Washington. https://cig. uw.edu/our-work/decision-support/, the section on Planning and Infrastructure. Full disclosure: your author is affiliated with them.

[4] If you are interested in water in the West, in the face of society's demands, a dated but still relevant work is M. Reisner, *Cadillac Desert*, 1986, Viking Press. The socioeconomic issues have not changed much, only their immediacy and severity, in the face of climate change.

[5] I have barely touched on water quality in previous chapters, for lack of space and expertise. For those interested in more coverage, one of my favorite places, a site of long-term water-quality research, is the Loch Vale Watershed in Rocky Mountain National Park. https://www2.nrel.colostate.edu/projects/lvws/research.html. See Figs. 2.7 and 8.1.

[6] Debris from fires? Usually not immediately as with the others, but dead trees can topple into streams, and decomposition of dead roots can undermine the stability of soils, causing rapid erosion.

[7] We have discussed that not all carbon in forests is stored in trees, but the tradeoffs that I will emphasize involve trees. Read on.

tradeoffs. For decades, society extracted timber from the Western Mountains, nearly at will, until its relative scarcity and other societal concerns[8] slowed things down. The need for carbon sequestration was not one of those concerns initially, but has steadily increased with better understanding of the global carbon cycle. The tradeoff between extraction and sequestration would seem simple intuitively: what is being removed isn't being sequestered. The complications come from considering different time *scales*. Ideally forests grow back after being cut down, either from planting or natural regeneration, if we ignore for a moment the issue of crossed thresholds (see Chapters 5, 6, and 8). Regrowth gains carbon back over time, and eventually the new forest might replace some or all of what was lost, or even more than what was lost. Here is where the complications set in. How long will it take, how much is gained or lost in the end, and what is the likelihood that the new forest will get to that point before perhaps being burned down, killed by insects, or blown over by wind? We have many models to project these outcomes, some even wrapping in other tradeoffs in the carbon balance such as the use of wood products to save us from the carbon cost of other structural materials. We have no universal answers, however, and even local answers are subject to the biases of models and modelers.[9] We lack these answers even for a stationary climate; projections into the future have the additional uncertainties from all the complexities we encountered in Chapters 4–6 and 8.

Air quality and visibility are services provided "for free" by ecosystems. They neither have to be extracted nor stored, and are noticed and appreciated most when they are degraded or lost. What degrades them can arise locally or from hundreds or even thousands of miles away. In the Western Mountains, poor air quality can have the same causes as poor water quality, such as waste from infrastructure (e.g., emissions from power plants) or resource extraction (effluent from mining). As I noted in Chapter 6, however, on the worst days for air quality the principal cause is fire. Even the smallest *prescribed fire* (fires set intentionally as part of management) can degrade local air quality so much[10] that over the years, concerns about air quality are the most frequent reason why prescribed fires that have been scheduled are canceled (Note: they would not be scheduled in the first place if the weather were hot, dry, or windy enough to cause concern about the fire's escaping) (Figs. 9.1 and 9.2). Days with the worst *regional haze* (widespread hazy skies) across the West are also caused

---

[8] The most visible being conservation in the Pacific Northwest (see discussions of the Northern Spotted Owl). Some of that, but as a colleague said once, "The timber industry didn't run into environmentalists, it ran into the Pacific Ocean."

[9] For example, many models predict the future carbon balance in forests with or without "management" (a euphemism here for repeated timber extraction). It has not gone un-noticed that the models built by those supported by the timber industry usually have different outcomes from those built by conservation groups. Can you guess which is which, whether management increases or decreases the carbon sink, from our previous discussions of the imperfect objectivity of scientific research?

[10] The worst smoke can be from small fires that are set by managers in order to reduce fuels on the ground. Logs, especially when they are rotten, smolder instead of truly flaming, causing much more smoke for the same amount of fuel consumed.

Fig. 9.1 El Capitan, Yosemite Valley. (**a**) A clear October day, (**b**) prescribed fire, also in October, set to reduce invasive plants in the understory. Photos by the author

Fig. 9.2 A burning stump releases huge amounts of smoke, in a prescribed fire to maintain white-bark pine. Photo by Robert Keane

by smoke from large wildfires that are intense enough to loft smoke into the atmosphere. So for air quality and fire there is a conundrum analogous to "pay me now or pay me later" (for the service).

Managers of public lands have many difficult choices, as we shall see in the next section. A major reason for setting prescribed fires is to reduce the severity of future wildfires, which may degrade air quality more than does the prescribed fire, which can be scheduled to minimize smoke downwind. Differences in the outcomes are hard to predict even in a stationary climate; the warming climate adds a big new uncertainty. The best approach may be to look at each situation afresh, but this is costly, of time and other resources, and *fragile* (i.e., doesn't translate very accurately) to extrapolation to other situations, even those that seem very similar.[11]

Some tradeoffs in ecosystem services have clearly defined stakeholders. For example, different people gain more, whether economically or otherwise, from timber extraction than from keeping a forest intact for hikers.[12] Sometimes the same

---

[11] What we wind up doing, as in other situations I have described, is to use models, wherein often we try to look at as general a setting as possible, in order to inform the broadest inferences possible. For example, if we look at one scenario in one watershed, and do experimental simulations in which we contrast frequent prescribed fire with no prescribed fire and look at air quality for the whole run of the experiment, one scenario may seem superior. But because of all the complexities involved, this result may not carry over into another watershed, or even a similar but not identical scenario for prescribed fire. Management is challenging.

[12] Although one could certainly argue that we all benefit from carbon sequestration.

stakeholders both gain and lose. People who build or move into the wildland-urban interface gain the benefits of open space, reduced congestion, and enviable views, all of which could be seen as ecosystem services, but lose by having direct threats to these, and their safety, from increased risk of wildfire.

Let us take a step back now and look at the social framework for providing and preserving ecosystem services from US public lands, which make up a large proportion of the Western Mountains. As we have seen, almost everywhere else is "downstream" from some public lands.

Mitigation, Adaptation, Management

I want to begin this section with humble kudos to those "on the ground" managing public lands in the Western Mountains. In the midst of political turmoil and rapidly changing directives coming from above, they balance many interests, often polarized and mutually hostile, with the goal of maintaining mountain ecosystems in the best possible state.[13] A warming climate makes this task more difficult, and a big part of it is understanding what you can or can't control or change.

Mitigation, in our context, is stopping or slowing what is going wrong. In climate change, that all starts with rising levels of CO_2 in the atmosphere, which is a global phenomenon because atmospheric CO_2 is *well mixed*. Managers' first responsibility, by design, is to their own domain, typically a National Park or monument, "multiple-use" areas that provide more ecosystem services beyond preservation,[14] or wilderness areas whose primary purpose is preservation but are within or adjacent to multiple-use areas. Recently, in light of increasing awareness of the warming climate, a global responsibility for mitigation has been added to managers' already full plate: carbon sequestration.[15] Their task is to maintain mountain ecosystems, especially forests, as carbon *sinks*. As we saw when discussing the carbon balance, different forces operate at different scales of space and time to determine whether a forest is a sink or a source. A forest that is undisturbed, by wildfire or insects or mass movements, or by logging, can be a slow sink for years, but then become a rapid source with one disturbance. Frequent disturbances, when smaller, mean that the changes from sink to source are smaller, and often more controllable (generally

---

[13] For what, or whom? And with respect to what? I'm being very general intentionally. There are many angles to this, and many stakeholders. Read on.

[14] I distinguish between *preservation* and *conservation* in this way. Preservation refers to a state, e.g., "pristine", whereas conservation refers to an element or a process, e.g., the wolverine or a free-flowing stream. Other use these words in many ways. As a reader, you won't need to hold to an exact definition here, unlike some other terms I have used.

[15] Of course, other types of mitigation could happen on public lands in the Western Mountains. We could shut down and replace (with carbon-neutral technology) major sources of fossil-fuel emissions, such as coal-fired power plants. But on-the-ground managers have little to say about that, and their work is our focus here.

a good thing). As with fire and air quality (and see Note 12), we have many imperfect but still useful methods for projecting the *source-sink dynamics* (i.e., the forces that determine sources and sinks). It is unfortunately not so simple as just letting all the trees grow.[16]

Adaptation is doing the best you can despite what is or will go wrong. This is a more central task for public-land managers in the Western Mountains than is mitigation, and it focuses on their local domain, with consequences downstream being secondary, though not ignored. Adaptation depends broadly on mitigation in that to do your best you have to have some idea of how bad things will be, and also specifically in that it is part of a larger framework for management that includes carbon sequestration. But adaptation has its own conceptual framework, from which actions arise.

A framework proposed by Western-Mountain researchers[17] has three stages: "resistance", "resilience", and "response". Resistance means stopping or forestalling whatever changes are imminent. Typically this would be highly valued resources that can't "move" out of the path of change. Two examples of these are the iconic glaciers in Glacier National Park and the giant sequoia groves in the Sierra Nevada. For the glaciers, we have seen (Chapter 4) that "resistance is futile". For the sequoias, a short-term strategy would be to *monitor* for drought-related stress and water each stressed tree individually.[18]

When resistance is not an option, resilience means increasing the range of climate under which an ecosystem (usually the scale of interest) can at least survive.[19] This means damping critical limiting factors or extremes that could be driven into intolerable ranges by warming climate. For example, in a water-limited forest, increased temperatures mean that each tree needs more incoming water to balance the greater loss through *evapotranspiration*. The more biomass in a forest, the more water it needs, in total. So reducing the total biomass, while preserving a functioning forest, makes it more resilient to climate change.[20] Another example, with different organisms, is resilience in animal populations from changed connectivity. For example, we saw (Chapter 7) that Canada lynx are at risk for local extirpation of their southern populations from a loss of critical habitat (snow cover). Decreasing

---

[16] And even if you could, there would be conflicts with other priorities given to managers.

[17] Specifically, in C.I. Millar *et al.* 2007. Climate change and forests of the future: managing in the face of uncertainty. *Ecological Applications* 17:2145–2151.

[18] I am not aware of any proposals actually to do this—we are not there yet—but drought stress is inevitable without significant global mitigation.

[19] Resilience has many definitions, both for ecosystems and in the world at large. They range from the arcane philosophical to the very mathematical. A good place to start, as with many concepts, is the Wikipedia page. https://en.wikipedia.org/wiki/Ecological_resilience. My definition links to our existing currency: limiting factors.

[20] Readers who are one step ahead might ask if it works the same in energy-limited systems. What we see there will more likely be loss due to indirect effects of climate rather than direct. For example, in alpine meadows (and most other cases) energy limitations will weaken, and previously energy-limited species, i.e., trees, will invade. See "response".

landscape resistance in its range could extend its ability to move such that it could reach areas that are snow-covered more easily.[21]

Resilience is finite. In our examples, eventually the forest will have to give up so many trees to survive drought that it will cease to be a forest, and the lynx will run out of snowy landscapes to support its prey, the snowshoe hare. Response means recognizing what you can't change and facilitating the transition of an ecosystem, or population, if possible, perhaps in a way that makes the transition less extreme.[22] For example, recall our "forests on the brink" in the Southwest or Sierra Nevada, and refer to Fig. 5.6. If that burned ponderosa-pine forest is no longer resilient to fire, and will succeed to a shrub field, maybe a few oaks judiciously transplanted from nearby lower elevations will keep some tree cover that is desired, maintaining some semblance of a forest. For animals, assisted dispersal (see Chapter 5, Note 31), to a suitable habitat with a climate like the one they have lost, may be an effective "response".

The Work of Adaptation

A systematic and structured process can facilitate adaptation to climate change throughout the Western Mountains, by being repeatable and self-correcting. We shall not dive deeply into the process here, but colleagues of mine[23] in the US Forest Service have developed a framework that serves well across very different types of ecosystems, from the dry Intermountain West to the rugged Cascade Range and Northern Rocky Mountains.

Identify Resources at Risk

For example, water resources and infrastructure, fisheries, vegetation, both forest and non-forest, wildlife, recreation, and ecosystem services. What will persist, or change (see Chapter 1)? What is the level of risk?

Where Do We Want to End Up?

Do we have a vision for what the landscape will look like, moving into the future? (usually not a static one). How much variability should we expect and accept? Should we try for the way it used to look, how the models say it will look, or just

---

[21] This is a tough one because of all the other factors limiting movement for the cat, but it is an example about how resilience can be relevant for vagile organisms, not just "ecosystems".

[22] Or a threshold or tipping point not so drastic.

[23] Led by David L. Peterson, Jessica Halofsky, and Linda Joyce.

how we want it to look?[24] Or should we aim for process (what persists) rather than specific states? A sobering constraint on any choice that relies on experience with other systems, whether they are historical or just elsewhere, is the "no-analog futures" paradigm. This means that given the current unprecedented rate of warming (see Chapter 3) and the many potential pathways for succession in complex ecosystems, returning to a historical state or duplicating the state of another system are both very unlikely. The future will bring a state with which we have no experience.

What Are Our Options?

Typically, there will be multiple *scenarios* considered. For each one: What do we do? How effective will it be? How soon is it needed? Where (on the landscape) will it work? Who can help? How much does it cost? What are the barriers?

Planning for adaptation includes all credible sources of knowledge. These are the "toolkits" I described in Chapter 8 ("How can we anticipate change?"). *Field-based knowledge* includes not only formal scientific research, indigenous knowledge, and anecdotal knowledge but also the many lifetimes-worth of first-hand acquaintance with the landscape of the on-the-ground managers of public lands, which is a mix of all the other types. Model-based knowledge comes from GCMs down to landscape disturbance models, tempered by the oft-healthy skepticism of the managers about the predictions of the latter.

Example, from the Northern Rocky Mountains

I have written about all our ranges with different filters at the ends of Chapters 3–8, and throughout Chapter 2. For a more detailed look at one range, through yet another filter, the challenges for adaptation to climate change, I draw on work[25] from those same colleagues, from the US Forest Service,[26] mentioned above. Within their framework, we can ask the three key questions.

---

[24] In management jargon, these correspond to "historical range or variability", "future range of variability" (I'm serious), and "desired future conditions".

[25] Specifically, J.E. Halofsky, *et al.* 2018. Climate change vulnerability and adaptation in the Northern Rocky Mountains. US Forest Service General Technical Report RMRS-GTR-374. Rocky Mountain Research Station, Ft. Collins, Colorado. Free online at https://www.fs.fed.us/rm/publications/titles/rmrs_gtr.html (scroll down a bit). Long, but quite readable if you don't try to do it in one session.

[26] Who is in charge of which public lands in the Western Mountains may seem needlessly opaque. National Parks are part of the Department of the Interior, which also runs the US Fish and Wildlife Service (creatures) and the Bureau of Land Management (non-forested lands, sometimes called "rangelands", reflecting a bias toward certain "services"). The Forest Service is in the Department of Agriculture (really!), and manages wilderness areas, forested or above treeline, that are not in national parks. The Forest Service is charged with adaptation on most forested landscapes within the domain, although work such as that cited in Note 25 brings in participants from different agencies, along with many other stakeholders.

What Resources are at Risk?

The principal categories are (1) water resources and infrastructure, (2) fisheries, (3) forest vegetation, (4) rangeland vegetation, (5) wildlife, and (6) recreation. Each faces multiple threats, many of which we have seen in previous chapters.[27] Threats to each entail possible losses of *ecosystem services*.

Loss of glaciers and snowpack will affect water resources, and timing of stream-flow will force changes to infrastructure, both physical and logistical. Rising stream temperatures will affect native trout species, notable bull trout and cutthroat trout. Forest vegetation will become more water-limited and stressed from drought, with consequences for cone and seed crops and regeneration (remember that this is the most vulnerable stage), possible losses of genetic diversity, and stress from ecological disturbances: wildfire, insect outbreaks, and pathogens. Rangeland vegetation will be affected by rising temperatures, with lower soil moisture as a result, and possible changes in *phenology*. The climate-related threat to some animal species is that a warmer climate may produce conditions that exceed their physiological limits. Other factors, such as loss of habitat or connectivity, diseases, including exotic ones, and human development, may interact with climate and push species across physiological thresholds. Different mammals, birds, and especially amphibians are vulnerable.[28] The threats to recreation are most obvious, and almost certainly greatest, for snow-based winter activities, which make up about 30% of national-forest visits at higher elevations. Loss of snow cover will affect both ski resorts and backcountry skiing. (Warm weather activities may actually increase, though perhaps at higher elevations, on average).

Where Do We Want to End Up?

Most of the objectives agreed upon[29] for these resources focus on the "resilience" and "response" stages of adaptation. Ecosystems and ecosystem services (e.g., glaciers) may be beyond the point at which "resistance" is possible. For example, several options are proposed for increasing resilience for whitebark pine, which faces multiple threats, as we have seen. If managers can increase resilience for this key species it may forestall a transition to vegetation without it ("response"). In contrast, if some element of an ecosystem that is critical for a species survival disappears, such as snow for Canada lynx or wolverine, it means the "response" stage. Vulnerable animals could be relocated, if there is available habitat elsewhere, and their local extirpation has to be accepted.

---

[27] Quiz: how many of these apply to all the Western Mountains, and how many are limited to some or just the Northern Rocky Mountains?

[28] See the reference in Note 25 for examples.

[29] Specifically, with the inputs from different stakeholders reported in the reference in Note 25.

What Are Our Options?

The final step of a strategy for adaptation is to recommend specific action. Continuing with our example of whitebark pine, possible steps include focusing on *refugia* (in this case, sites whose *microclimate* may be buffered from warming) for active management, using prescribed fire to reduce competition or fire severity, and actually planting trees at lower elevations, preferably from genetic stock that is resistant to the blister rust (see Chapter 6) (Figs. 9.3 and 9.4). Taking a step back from a single species to the whole landscape, we need to consider multiple spatial and temporal *scales* in strategies for adaptation. This is important both logistically and practically. First, actions may interact, possibly neutralizing each other or having cumulative effects that we need to anticipate. For example, tactics recommended to promote resilient communities of whitebark pine and quaking aspen (another species of concern) involve prescribed fire to restore historical fire regimes, but these were two different fire regimes: moderate severity for whitebark pine and high severity for aspen. Where the species overlap in range one or the other will be a loser. Moreover, the cumulative effect of these actions will be to support fire regimes unsuited to the "non-analog" future of a warming climate, being relics of a cooler past.

The practical issue with large-scale action to promote resilience (or ease a transition, i.e., "response") is that it is costly, both of time and personnel. With finite

Fig. 9.3 Prescribed fire in whitebark pine stands to restore an open structure and make it more resilient against the blister rust. Photo by Robert Keane

Fig. 9.4 Whitebark pine has dispersed downslope (counter-intuitively in a warming climate), because of fire exclusion at the lower elevations. This dispersal can be mimicked by planting. Photo by Robert Keane

resources, there will always be unpleasant but necessary choices. Consequently a big part of planning for adaptation is setting priorities. How likely is something to be lost? How bad will that be? How much can we save, and for how long? Finally, how do we know if it's working? *Monitoring* (see Chapter 8) is under-appreciated and usually underfunded as a less glorious part of science, but a complex process can stray from expectations quickly, so we need to watch as closely as we can, and expect the unexpected.

Surprises

That "unexpected" can arise in different ways. It could be just miscalculation, where we had too much confidence in a projection.[30] It could be the equivalent of the proverbial "snake eyes", where our outcome was something very unlikely, but will still

---

[30]This could be literally the equivalent of a mistake in algebra, where we relied, for example, on a computer program with a bug. Surprisingly often, however, it could be *false precision*, where we thought we could calculate a more detailed answer than was warranted. Recall our example (Chapter 4) of predicting the high temperature in one location on April 15, 2040.

happen once in a while.[31] We are wont to call these "extremes" when they are unwelcome events, such as heat waves, and "anomalies" when they are relief, such as a cool wet summer in the desert.[32]

In a situation with many possibilities of roughly equal likelihood, any outcome can be expected, and therefore "unexpected" because its probability of occurring is small. This could mean that our *models* (conceptual and other types) of the system are simply inadequate. For example, suppose that we are concerned about a mountain lion population (see Chapter 7) that might be very inbred, and if it drops below 50 animals it is likely to be extirpated locally. Our simulation model[33] projects that in 10 years, the population will be between 15 and 250. With this uncertainty, we cannot "expect" either success or failure in preserving the population.

How important is it not to be wrong? Surprises can be inconsequential, or game-changing. Recall our discussion of ecological thresholds in Chapter 8, and the importance of knowing both how big a "move" your ecosystem may make and how close it is to the threshold. So a move that is unexpectedly large, but not close to a threshold, may be less consequential than a move that is just a bit larger than we expected, but crosses the threshold. For example, we can compare two forests, one energy-limited and the other water-limited. In the first, suppose that our projection of average summer temperature for the next decade proves to be low by 3 °F (yes, that's a lot!), but for the second, it is low by only 1 °F. In the first, we will see more growth of trees, and maybe some shifting proportions of species, but it will still be a forest.[34] In the second, we could see a threshold crossed to non-forest, for example, after a disturbance that kills most trees and when seedlings cannot tolerate the warmer temperatures.

---

[31] There are 36 different ways (6 for the first die * 6 for the second) that two dice can display when rolled, so the chance of snake eyes (1,1) is 1 in 36. But if ("ecological") dice are being rolled a lot, say 1000 times a day, then about 28 times per day, on average, that very unlikely event will occur. Given the number of annual, or decadal, events in the Western Mountains, even a one-in-a-thousand outcome will happen occasionally.

[32] James Hansen aptly compares changing temperature extremes in a warming climate to loaded dice. What if your dice were engineered so that it was almost impossible to roll a 6, and pretty hard to roll a 5? You will see a lot more snake eyes (hot weather). If we are running *stochastic* models, one way to evaluate them is whether or not the frequency of anomalies matches that in the real world. This is just as important as getting the averages right.

[33] Recall that I said in Chapter 7 that animal population models abound. *Population viability analysis* (PVA) uses these models and other tools to predict whether local or global species will survive, and for how long.

[34] But the alert reader will notice that I am ignoring other "surprises". Perhaps the hotter weather will make for an unexpectedly severe fire, or bark beetles may reproduce more rapidly than expected because of the warmer weather. Even so, will the consequences be as great as a transition to non-forest?

"Anti-fragile" Management

What if there were a way to roll with all the surprises, or even better, to take advantage of them? Adaptation in the management of public lands, as the word suggests, happens in response to change. When change is unprecedentedly fast, as is the current warming of Earth's climate, we expect that adaptation might fall behind. *Resistance* may come too late, *resilience* may have too many pieces,[35] and *response* may be limited to "Wait and see what happens".

It has been proposed[36] that in different types of systems, including the ecological, economic, and social, some processes, structures, and institutions thrive at levels of disorder, or non-stationarity, that are above the minimal levels that one might wish for. In particular, there may be ways for management to "hitchhike" with rapid changes, or crossed thresholds, caused by extremes. Sometimes this could promote *resilience*, but probably more often it would shape a more robust *response*. For example, recall from Chapter 6 our discussion of wildfires in a warming climate, and of "patches", the relatively homogeneous "units" of a landscape. Where larger ecological disturbances are projected (much of the Western Mountains), we expect a coarser "grain" of the landscape: larger patches, on average, whether they be where a disturbance occurred or what was left over from it. So extremes, such as a massive wildfire or insect outbreak, provide a window into landscape structure of the future, and they can be focal points for active management such as planting species that are better adapted to a warmer climate or for *monitoring*.[37]

Opportunistic management is not a new idea, but it has happened more often in response to serendipity than to disorder or unpleasant surprises. We can use both, of course, but which is more likely in a rapidly changing environment?[38] Most of the good fortune may fall to the "winners" that we referred to in earlier Chapters 6 and 8, such as bark beetles and disease-carrying insects. But rather than stopping on that unpleasant forecast, let me reaffirm, in response, my admiration for the

---

[35] Recall our whitebark pine example, where we want to make populations of the species resilient to (at least) three types of stress: (1) blister rust, (2) encroachment by other species, and (3) more intense fire. Each of these suggests difference actions to produce resilience. (1) Finding genetic stock that is resistant, (2) using prescribed fire to maintain open stands, (3) but restricting condition under which prescribed fires are started or wildfires allowed to burn. With rates of change that were more "forgiving", there might be more time to implement all three methods strategically.

[36] In a book. N.N. Taleb, 2012. *Anti-fragile: Things that gain from Disorder*. Random House, NY. Hence the title of this section. Taleb is probably best known for claiming, with mathematical justification, that extreme events, sometimes called "black swans", are inherently unpredictable. This may seem like a dead end for management, but what if we could make the proverbial lemonade out of lemons by anticipating them? Read on.

[37] Which is often limited by finite resources and by being a (undeservedly) low priority. See Chapter 8.

[38] Of necessity, we are getting away from trying to keep things the same, or within the range of past variability, but that paradigm is hard to give up, especially for those of us for whom the historical state of the Western Mountains has been such a force in our lives. See next (and last!) section.

on-the-ground managers of public lands in the Western Mountains and my faith that they will continue to adapt in the years ahead.

Whither Wildness?

In wildness is the preservation of the World.
    —Henry David Thoreau

The Western Mountains have drawn people, either to live or to visit, for many reasons. The aesthetic dimension has probably always been a part of it, from the first Americans who arrived more than 13,000 year BP, through John Muir and his lesser known contemporaries, to the present. The Wilderness Act of 1964 and the Endangered Species Act of 1973 were societal and political responses to a growing interest in preserving aspects of the natural world that are exemplified by Western-Mountain landscapes.[39]

A bit later,[40] the discipline of *conservation biology* focused research into ecological changes that put species and other elements of ecosystems at risk, and lay-audience journals such as Wild Earth[41] focused on reporting of science, environmental activism such as proposals for wilderness areas, and aesthetic elements of natural areas, particularly large unmodified (by humans) landscapes in designated wilderness or national parks.

The last decades of the twentieth century saw voluminous philosophical discussion of the idea of wilderness (or wildness, *a la* Thoreau), ranging from full-fledged advocacy of the wilderness experience as salvation for mechanized society to "deconstruction" of what is considered a spurious notion. Mastering and parsing these arguments is beyond my expertise,[42] and from what I do understand, there is

---

[39] Not all US wilderness areas or endangered species are in the West (especially if you exclude Alaska from the "West"). Of the 803 designated wilderness areas in the US, the five largest, and 14 of the 20 larger than 1,000,000 acres, are in Alaska. But only one over 1,000,000 acres is outside the West or Alaska, the Boundary Waters Canoe Area in Minnesota (19th largest). Unsurprisingly, Hawaii is the state with the most endangered species (503), followed by California (299), Florida (135), Alabama (122), and Texas (103). In general, the farther south the state, the more endangered species it has. Endangered fish (79 species) outnumber mammals (37 species) and other animal life forms (birds, reptiles, insects).

[40] The first issue of the journal *Conservation Biology* was in 1980. Topics addressed in its early years included population viability, landscape fragmentation, *keystone species*, problems with inbreeding, and the effects of human activities such as clearcut logging on all the above.

[41] *Wild Earth* ran from 1990–2004. Its publishers had left the Earth First! movement, unhappy with that group's direction.

[42] And I confess beyond anything but a passing interest. The light-to-heat ratio of much of the discussion does not seem particularly favorable. But don't take my word for it. A strong counter-argument to the idea of wilderness comes from William Cronon, at https://www.williamcronon. net/writing/Trouble_with_Wilderness_Main.html, or in more detail in his books. The "pro" is best experienced live, in my view, by getting "out there".

not that much to inform our particular question, which is "What will happen to all this in a warming climate?"

I believe that we can frame the future of wildness in the Western Mountains in terms of our phrase from Chapter 1: "what persists, what changes". Each of us could specify tangible things that we associate with wildness. Here are mine (1) Clean air and superb visibility, (2) remoteness,[43] (3) danger ("This is not Disneyland"),[44] (4) no management, development, or resource extraction, (5) big fierce animals, (6) topography, (7) "natural wonders". I hope that if you have come this far with me in the book you should have a good feeling for how to apply the same reasoning to yours. Will they persist, or how will they change?

Clean Air and Superb Visibility

The worst air quality and least visibility in the Western Mountains happen mostly on days with fires burning upwind that are intense enough for smoke to rise into the prevailing winds. These can be prescribed fires as well as wildfires. Smoke in the Western Mountains is very likely to increase in a warming future, though not everywhere and not all the time.[45] Smoky as some days seem, we may be living in a sweet spot for visibility right now in the history of human inhabitation of the West. There is good evidence that in much of the West, Native Americans may have used enough fire, of enough intensity, to produce as much or more smoke overall[46] as do sporadic wildfires and prescribed fires that are set on days when air quality is forecast to be affected minimally. Air quality and visibility will change, almost certainly for the worse.

---

[43] How far is far? Remote medical providers are taught that if you are an hour or more from "definitive medical care", you are in a "remote" setting and can provide medical services that are outside your purview in other settings. For example, you can attempt to *reduce* (basically straighten) a major leg fracture, to save a limb from loss of circulation. But if you are miles from the nearest road, but have cell service and can call a helicopter to evacuate a patient, that's not "remote". Likewise, how remote does it feel to be on a 14,000 ft. summit when fellow summiteers are chatting away on their cellphones?

[44] More specifically, you can make mistakes that might not get you into trouble in the city, or even on the farm, but will do so in the *wild*. "Your safety is not guaranteed."

[45] Recall our discussions of variability and of fire regimes that are flammability-limited vs. fuel-limited.

[46] A legitimate argument against this is that current prescribed fires produce a lot of smoke because of their type of combustion (smoldering), due to the fuels' being largely dead wood, often large pieces that burn "dirty". Native Americans burned mostly live vegetation, having different (a wider range of) objectives from present-day fire managers. We will never know for sure, but the idea of crystal-clear air across the West before Euro-American conquest is a myth.

Remoteness

We will have much more choice[47] about this quality, and the next three, than about air quality and visibility. The biggest threats to remoteness are digital and demographic, not climatic. As I write, many areas in the West are still without cell-phone service, but these do not correspond particularly well with areas that we might consider remote. Cell phones don't work on country roads, but they work on some mountaintops in the Sierra Nevada, the Rocky Mountain Front Range, and probably many other summits.[48]

With a growing population, and more housing developments abutting wilderness areas, we might expect that on average there will be more people everywhere in the "wild", with more cell phones and less seclusion. In the last 10 years, visits to national parks have increased about 15%, but most of this has been in the famous parks, parks near urban centers, and parks recently designated.[49] So if it were just affected by climate, remoteness would seem likely to persist, but other factors take predictions out of the hands of those of us who focus on warming climate. The most remote areas, in particular, seem unlikely to be domesticated soon.[50]

Danger (or the Need for Self-Reliance)

There are two pieces to this one (1) Will wild areas become more or less dangerous, and (2) If you are unlucky, are you more or less likely to be saved? There is good reason to believe that in response to warming, *objective danger* (adventurers' lingo for things you can't control) in the wild will persist and may even increase. For example, in the world of rock and ice, stability, and therefore cold, is good. In the Western Mountains, and around the world,[51] rising temperatures have destabilized

---

[47] Assuming that we don't "choose" to *mitigate* future warming! As I write, the chances of doing that, even not nearly enough, seem slim.

[48] Thankfully, I have experienced this personally only on the first two.

[49] See   https://www.nationalgeographic.com/travel/destinations/north-america/united-states/national-parks/avoid-overtourism-indiana-dunes-gateway-arch/ and sources therein. The good news for remoteness advocates is that in the big-name parks, such as Yosemite and Yellowstone, visitors tend to cluster around the tourist attractions rather than fanning out into more remote areas. As our population continues to grow older on average, we are unlikely to see a big increase in backcountry travel soon.

[50] As a conscious backlash to all this, Dave Foreman, the founder of Earth First!, proposed the creation of "super" wild areas. Therein there would be no communications except in person, and no rescues. ("Don't leave notice of where you're going with anyone"). This idea never went anywhere publicly, except that some intrepid individuals practice it for themselves and their companions, i.e., intentionally eschewing any outside support.

[51] For example, in 2012 expedition leaders at Mt. Everest canceled ascents that were already underway, citing the worst instability in ice and rock formations that they had seen. (Sorry clients, no refunds).

terrain dominated by snow and ice (Fig. 9.5). For winter travelers in the backcountry, variable temperatures, along with more precipitation, especially *rain-on-snow*, create more threat of avalanches, as we saw in Chapters 4 and 6. In the world of dense forests in rugged topography in roadless areas, the greater chance of severe wildfires in a warming climate[52] increases the chances that summer travelers will be trapped.

The second piece of this one, whether exposure to danger in the wild will be less lethal, depends on factors unrelated to climate. For closure, I shall leave these predictions to others but note that danger from other humans is a significant part of deciding what precautions to take in wild areas. Others in one's party or in intertwined parties above or below you on the same mountaineering route can make mistakes. Some will have bad intentions,[53] just as anywhere else. Such risks will persist.

No Management, Development, or Resource Extraction

Of these, management is the issue that depends sensitively on the effects of climate change, both observed and expected. Even in a stationary climate, preserving wild areas requires some types of management. Limiting visitor use[54] has become critical to preserving wildness, but active management in the field is also critical. For example, fires are set in giant sequoia groves in Sequoia National Park, and in many parts of Yosemite National Park, to reduce fuels that could burn intensely enough to kill the sequoias (in the groves) or other trees that might not regenerate after fire in a warming climate (Fig. 9.1). Another type of management that has sparked more serious debates about "naturalness" or "wildness" than prescribed fire is the *assisted dispersal* that I discussed in Chapter 7. This is a "response", in terms of adaptation, but detractors argue that creates "anthropogenic" ecosystems, not those that would have evolved absent of a human influence. Can wildness persist, even with these managed changes?

---

[52] Because these are *flammability-limited* fire regimes, where we expect wildfire extent and severity to increase. See Chapter 6.

[53] Not helped by regulations that promote reckless behavior. Perhaps it is the delusion that "guns make people safer", but the recent decision to make it legal to carry guns in US national parks will surely increase danger levels, and make "wildness" more like "wild West".

[54] One day in the late 1960s I was hiking the Mt. Whitney trail and was approached by a field ranger who asked to see my wilderness permit. I had never heard of such a thing, and of course didn't have one, but it would have been a formality back then. Over the years, the supply/demand ratio has shrunk so that as many readers must know, when popular campground sites open for a particular day, it takes 10–15 s for internet requests for them to take all the vacancies. This is also starting to happen for popular trailheads.

Fig. 9.5 The north face of Mt. Kennedy in the Yukon (yes, not in our domain but the best photo I know). Already dangerous, this mixed rock and ice will likely destabilize in a warming climate (in the Arctic warming roughly twice as fast as in the West). Photo by Lara-Karena Kellogg

Big Fierce Animals

In the mountains, your head changes when something can kill you. It could be a slip or an avalanche when on a climbing rope, or a bear, lion, or rattlesnake when simply walking. This is danger, discussed above, but also, for many of us, respect for other creatures,[55] and a sense of wildness from not being in control. Our big fierce animals in the Western Mountains are *habitat generalists*, and secure right now from global extinction, but they are still at risk for local extirpations, primarily from loss of habitat and *connectivity*, as we saw in Chapter 7 for the mountain lion, but also indirectly from climate change. For example, grizzly bears in the Greater Yellowstone Ecosystem (GYE) consume nuts from whitebark pine, which are a high-calorie food suited to gaining weight prior to hibernation. If whitebark pine were to succumb to the double threat of bark beetles (abetted by rising temperatures) and blister rust, it would produce another stress for the already stressed bear population in the GYE. Even if claims about their recovery are true, which has been challenged,[56] their numbers are still barely above those of a *minimal viable population* (literally, just enough animals to avoid extirpation from genetic or environmental causes).

Topography

Because our mountain landforms change in *geological time*, the current rapid climate change will leave their main structural features intact. The only exception is that topographic features that are partly or wholly supported by ice could become unstable and disintegrate. I know of no specific locations in the Western Mountains that are at risk, although more intrepid mountaineers than I may be aware of structures that are. This is an issue elsewhere in the world.[57]

"Natural Wonders"

The "Seven Wonders of the World" were a fixture in the children's geography books of my generation. Back then only ancient ones were listed, but more recently many other lists have been proposed, including seven "natural" wonders.[58] One of these is the Grand Canyon, so starting with that let us ask specifically what might happen

---

[55] Walking a few years ago in Banff National Park, BC, Canada, my wife and I encountered a warning sign: "Caution, sow with cubs seen here" (meaning: Protective and potentially aggressive mother grizzly bear is nearby. She could outrun you and kill you with one swipe of the paw). We decided not to hike in grizzly-bear habitat any more. Fear and respect both, and definitely a sense of wildness.

[56] See D.F. Doak and K. Cutler. 2013. Re-evaluating evidence for past population trends and predicted dynamics of Yellowstone grizzly bears. *Conservation Letters* doi: 10.1111/conl.12048. Free online (as I write) at https://doaklab.org/.

[57] There is (was!) a famous section high up on Mt. Everest on the route by which Tenzing and Hillary made the first ascent in 1953. This was a steep 40-ft. pitch called the "Hillary Step", mostly rock but (evidently) secured to the mountain partly by ice. It was most likely shaken loose by the 2015 earthquake, but very possibly destabilized before that by loss of ice.

[58] See https://en.wikipedia.org/wiki/Wonders_of_the_World for an overview.

with some natural "wild" wonders of the Western Mountains. As with the structure of this whole section, I will propose a few, and encourage the reader to apply the same thinking to her or his favorites, informed (I hope) by this and other reading.

(1) The Grand Canyon, (2) Yosemite Valley, (3) Utah Red Rock Wilderness, (4) The Olympic Rainforest, (5) Yellowstone, (6) Mt. Rainier and the other Cascade volcanoes, (7) alpine meadows (in most of the ranges). My choices vary with respect to how important is topography vs. other more ephemeral features of wildness.

The Grand Canyon

The Grand Canyon is certainly about topography, but also water. With a changing hydroclimate in the Southwest, and increasing demands on the Colorado River, we can expect changes in amount and seasonality of runoff. This will not only affect travel on the main river, but also change water levels in the side canyons and access to them.[59] So despite its structure remaining mostly intact, the Grand Canyon's wildness, from the perspective of visitors, could change.

Yosemite Valley (and the Rest of the Park)

The experience of wildness is complex in Yosemite, because more than 90% of visitors stay in the Valley, less than 1% of the area of the Park. The Valley is perhaps the most iconic part but also the most developed. Its granite landforms change in geological time, even more slowly than the Grand Canyon, but as in the Grand Canyon, the rivers and waterfalls of the larger Park respond to the hydroclimate. But the biggest threat to Yosemite's character from a warming climate is to the vegetation: wildfires followed by failed regeneration by trees and a change from forests to shrublands (as I mentioned in Chapters 5 and 6 for the Sierra Nevada). The giant sequoia groves may survive for a while because of their location in *cold air drainages* and the trees' resistance to fire, but other low-elevation forests, including those in the Valley,[60] are vulnerable. With increased wildfire, both in the Park and upwind across California, reduced air quality and visibility from smoke will likely be more frequent and severe.

---

[59] Seasoned floaters on the main river consider that the many steep side canyons have most of the wildness in the Grand Canyon.

[60] As with other developed areas, Yosemite Valley will see major efforts to suppress wildfires and reduce their severity with prescribed fires. This should help for a while.

Utah Red Rock Wilderness

Distinctive landforms give the character to the five national parks in southeastern Utah and to much of the surrounding landscape, but warming climate may bring big changes to their less permanent features, both directly and indirectly. With hotter temperatures, less rainfall, and the ensuing drought, major dust storms could increase, impairing visibility and making cross-country travel difficult (Fig. 9.6). Dust from these storms can fall on snow-covered landscapes, such as the San Juan Mountains, directly downwind in southwestern Colorado, decreasing the *surface albedo* (as we saw in Chapter 3), thereby accelerating global warming. The combination of heat, wildfire, and the presence of exotic plant species could mean a future with more wildfire in this *fuel-limited* ecosystem, with continuous fuel in the form of exotic grasses connecting patches of vulnerable, but formerly sparse, native vegetation.

Fig. 9.6 Dust storm in the Utah canyon country. Photo by Ray Bloxham, courtesy of the Southern Utah Wilderness Alliance. https://suwa.org/

The Olympic Rainforest

Large tropical rainforests exist in the lower latitudes,[61] but temperate rainforests are limited to the Pacific coast of North America. The southernmost temperate rainforest that is largely intact is in Olympic National Park, in northwestern Washington. This low-elevation forest is an energy-limited system, and the maritime climate, here and on the western side of the Cascade Range, provides more of a buffer against rising temperatures worldwide than in other Western Mountains. The chief threat to these forests in coming decades would be from increased wildfire, but we have unusually poor projections, compared to those for the rest of the West, of changes in the fire regime. In the past, large wildfires have been carried by anomalous dry east winds. Then and now, these forests are 100% flammability-limited, so that it takes a "perfect storm" of weather to permit a large wildfire, and we don't have good regional forecasts of such future perfect storms as to whether there will be more or fewer.

Yellowstone (and the Greater Yellowstone Ecosystem (GYE))

The geysers on the Yellowstone Plateau are the biggest tourist attraction in the Park, but the forests of the plateau and the steep mountains of the GYE, including the Tetons and the Wind River Range, provide some of the wildest and remotest terrain in the Western Mountains. Both of these features may change in a warming climate. Though a *geothermal* phenomenon (literally, heat from the Earth), the 400–500 geysers also depend on water from snowmelt—thus a *hydrothermal* system—so warming air temperatures and drought could reduce their activity. On the other hand, variability over time in water temperatures, associated with uplifts belowground, can do the opposite: increase temperatures so that only steam erupts from geysers.[62]

Some of the best of our *inverse* models of climate and wildfire project that the average forest area burned in the Northern Rocky Mountains will increase substantially in warmer climate.[63] In this *flammability-limited* ecosystem, a shorter interval between fires (a straightforward calculation from the greater area burned) could mean that many trees of the dominant species, lodgepole pine, could be killed by fire before they are mature enough to produce *serotinous* cones, thereby preventing

---

[61] But less and less every year. In 2018, they covered about 6% of Earth's land area, down from 14% formerly. Almost all of the loss is from clearing by humans.

[62] This happened between 2003–2006, and although some toxins were produced that killed animals (bison), geologists tell us that there is no immediate threat of a volcanic eruption. Here is a reminder, though, that not all geological processes require geological time scales.

[63] Specifically, from greater *water-balance deficit* associated with warmer temperatures and the same available moisture from rain and snow fall. See Chapter 6, and Chapter 8 regarding a "tipping point" past which forest could become shrubland.

regeneration. Lacking seedlings of lodgepole pine, what were once forests could change to shrubs.

The grizzly bears in the GYE are the southernmost population, and isolated from those farther north. Though said officially to be recovering (but see Note 56), they are still at risk not only from loss of habitat, but also from potential loss of a principal food source, whitebark pine nuts, if that tree species is lost or greatly reduced in the GYE. Though arguably not directly at risk from climate change, the grizzly could be a victim of multiple factors, some exacerbated by warming climate.

Mt. Rainier and Cascade Volcanoes

These signature elements of the Pacific "Ring of Fire", along with the rest of "America's Alps", provide many of our elements of wildness. Changes to the hydroclimate, though muted by the maritime influence in the Pacific Northwest, could still affect the extensive *cryosphere* therein. While lower elevations are losing snow cover, higher elevations, including the flanks of the volcanoes and some remote and challenging summits,[64] may see instabilities from melting ice. Backcountry skiers and snowboarders may encounter more complex and challenging conditions, from snowfall in more varying amounts and with more varying textures and stabilities.

For the more risk-averse (most of us!), the sequence of ecosystems on the slopes of the volcanoes, from continuous forest to parkland to alpine to rock and ice, will be under "pressure" to move upslope. Barring major disturbance, however, this movement will lag behind climate change, as we saw in Chapter 5. At lower elevations, forests provide a *microclimate* that is less variable than what occurs without tree cover, so that mature trees can survive in warmer climates. At upper elevations, plants will be slow to colonize newly bare (from loss of permanent snow) terrain. On the Cascade volcanoes and elsewhere in the West, upper treelines, along with highly valued alpine meadows, will see complex changes.

Alpine Meadows

Mountain enthusiasts of all types know that a lot of life goes on between the highest-elevation trees and the upper end of habitat for plants. Alpine meadows lie under snow for most of the year, then flourish briefly, with hardy grasses and seemingly fragile wildflowers. Many of these areas are protected, in national parks or designated wilderness, because of their beauty but also because the potential for resource extraction is limited, unlike timbered areas lower down. The highest ranges in the West have upper treelines, and "alpine" ecosystems above them. We saw in Chapter 6 that two opposing forces will affect the treeline *ecotone* in a warming West: trees

---

[64] For example, the Picket (sub)Range, east of Mt. Baker and perhaps the most rugged area of the Cascade Range, with 21 peaks over 7500 ft.

invading slopes above them and increased wildfire killing some of those trees along with more trees in the forests below treeline. Consequently, ecotones will become more complex and wider, with some of our alpine meadows becoming a mix of flower fields and isolated trees or "tree islands".[65] This will make the visitor experience less "pure" (i.e., fewer pure alpine meadows), but the ecotone will be wider and more varied.

What Next for those Who Love the Western Mountains?

As promised, what you have read here is neither a polemic nor a guidebook. It is a very selected ecological narrative of the past, present, and future of the Western Mountains. What should we do with the information herein? I say "we" because I would wager that I have learned as much from the process of telling the tale as I expect that anyone will from reading it. Here is a multiple-choice answer to the question, with a bit of elaboration on each choice: (a) mitigate, (b) adapt, (c) educate, (d) experience, (e) all of the above. Depending on your age, education, work and life experience, and motivation, you could take different paths.[66]

Mitigation

Mitigators work to reduce the global footprint of greenhouse emissions, seeking to slow global warming and its effects.[67] Those in positions of power can work directly on that, and advocates can work on those in power to do so. Those with a technical or modeling background can engineer products or protocols to reduce the footprint. Regulators can approve products or enforce protocols. Current or prospective students can choose from a variety of careers, including STEM, social science, law, and international relations. Writers, performing artists, and visual artists can give us stories, songs, and pictures.

---

[65] Literally, these are clumps of trees, sometimes *clonal* (multiple trees from the same root stock). In islands, trees protect each other a bit from the elements, so there is survival in numbers at the extremes of their ranges.

[66] Including an empty path, i.e., not changing anything you do in the world. I am not necessarily advocating any of these paths, nor calling out anyone for not taking them. I will, however, register my disapproval for those who actively obstruct one or more of them.

[67] And eventually even reverse it, but as I have said, there is already a warming "debt", thanks to the long life of CO_2 in the atmosphere.

Adaptation

Adapters work to minimize the effects of the global footprint. In the Western Mountains, the leaders of this work are the aforementioned field-level managers on our public lands. Many of these folks are mid-career or later; current or prospective students working in climatology, hydrology, ecology, social science, economics, management, biochemistry, physics, and engineering could take their place in the future. People with expertise in medicine, including mental and public health, can help residents (of the West and elsewhere) be resilient and respond to the inevitable changes ahead. Writers, performing artists, and visual artists can give us stories, songs, and pictures.

Education

Educators illuminate the what, the how, and the why: detection, attribution, and explanation. Some learners will want to stop at the first step,[68] whereas others will take a deep dive. All three steps have value. We can learn about the Western Mountains in the field, at the computer, and via all the media "in between" them. Educators can use any or all of those media, depending on their skills, interests, and mobility and those of their students.

Experience

Go to the Western Mountains, or any mountains, however you can. They are not Disneyland, so go with care if you are going in person. Some places are more accessible physically than others,[69] and some have to control both the number of visitors and their activities. The climate is changing, and the effects of its change will be felt, but most Western Mountain ecosystems will *persist* for most of the lifetimes of most readers of this book, albeit with the changes about which you have just finished reading (Fig. 9.7).

> *May the road rise up to meet you*
> *May the wind be ever at your back*
> *May the sun shine warm upon your face*
> *May the rain fall softly on your fields*
> —from a traditional Irish Blessing

---

[68] Or before, particularly with topics for which even detection conflicts too much with prejudices.

[69] For example, as I write only one person has climbed El Capitan, Yosemite's 3000-ft. high granite monolith, unroped and alone, about 5000 people have "through-hiked" the 2650-mile Pacific Crest Trail (but fewer than 100 more than once), millions have visited Yosemite and Yellowstone National Parks, and billions can view their scenery on the internet.

Fig. 9.7 Where wildness remains. These are two of my favorites. Choose and visit your own. (**a**) The Cathedral range in the Yosemite high country, photo by Doug McKenzie, (**b**) Ambush Peak(s) in the Wind River range, photo by Jeff Whidden

Glossary[1]

Accumulation vs ablation zone Referring to glaciers, the accumulation zone is the upper part, where snow and ice increase. The ablation zone is where they decrease.

Adaptive seasonality The adjustment of an insect's reproductive cycle to coincide with the seasons, so that breeding happens at the same (optimal) time every year.

Adaptive vs evasive strategies Referring to how tree species survive wildfire. Adaptations include thick bark and *serotinous* cones; evasion is simply not being there when fire strikes.

Adjustable parameter In a mathematical model, some parameters are fixed for good, such as the "2" in the formula for the area of a circle. Others ("adjustable") can change based on context.

Aerosol A suspension of fine solid particles or liquid droplets in air or another gas. Aerosol concentrations in the atmosphere are the largest uncertainty in model-based estimates of future global temperatures.

Age structure The distribution of ages in populations of plants (ususally trees) or animals.

Atmospheric rivers Concentrated flows of moisture in the atmosphere.

Biotic vs abiotic Referring to living vs. non-living. An insect outbreak is a biotic disturbance, whereas wildfire is an abiotic disturbance.

Blister rust A fungus causing disease in five-needle pines, particularly whitebark pine. The fungus that attacks this species originated in China.

Boreal vs austral or tropical "Boreal" vs "austral" refers to aspects of the Northern vs. the Southern Hemisphere, e.g., "boreal summer", beginning June 21, vs. "austral summer", beginning December 21. In contrast to "tropical", boreal refers to aspects of the higher latitudes, e.g., "boreal forest".

---

[1] Italicized terms in entries are entries themselves.

© Springer Nature Switzerland AG 2020
D. McKenzie, *Mountains in the Greenhouse*,
https://doi.org/10.1007/978-3-030-42432-9

Boundary conditions Certain constraints from a larger-scale model that are applied to a finer-scale model. Typically in climate-change research these will be outputs or averages from a GCM that are used to keep a regional climate model, vegetation model, or air-quality model from "wandering away" from average conditions.

Broadleaf vs. needleleaf A type of plant, usually a tree, with "real" leaves instead of needles.

Capacitor A mechanism for "holding" energy or its equivalent for future use. For example, in our use, snow and ice are capacitors that hold flowing water for when it is needed.

Carbon balance Measurement of the additions and subtractions of carbon (in any form, i.e., inorganic or organic) to and from an ecosystem of any size, up to and including the whole Earth.

Climate sensitivity The change in global average temperature from a doubling of the atmospheric concentration of CO_2. The estimate that is generally accepted is ~3 °C, but this may change as CO_2 levels rise.

Climate velocity This is climate-change jargon for the rate (e.g., kilometers per year) and direction (toward the poles) that key metrics of climate move latitudinally as the Earth warms. It is used mostly to calculate whether species' dispersal could "keep up" with the climate to which they are adapted.

CO_2 fertilization The (presumably positive) effect that increasing atmospheric CO_2 has on plant growth.

Cold-air drainage Movement of colder air downslope (i.e., to lower elevations), displacing warmer air, a type of temperature inversion found in narrow valleys and canyons.

Complacent In our context, controlled little or not at all by limiting factors. The expression is commonly used in *dendrochronology* to describe trees whose growth rates do not seem to be controlled by climate.

Coniferous Trees with cones.

Connectivity A measure of how permeable a landscape is to the movement of organisms. It could be connected for some organisms and fragmented for others.

Convective storms Storms, such as thunderstorms and tornadoes, with strong movement of moist air in the atmosphere.

Core vs peripheral habitat The most important vs. less important habitat for an animal. These are often concentric, with the core enclosed, but not always.

Coupled models Two or more independent models that are integrated in a multi-disciplinary study. GCMs are often coupled with models of ocean dynamics, and can be coupled further with vegetation models and land-atmosphere exchange models into *Earth-system models*.

Cryosphere Anywhere that water is in solid form, whether snow, ice, or permafrost.

Defoliators vs cambium-feeders Two types of insects whose outbreaks can kill trees. Defoliators eat leaves or needles, and if they eat enough photosynthesis is no longer possible and the tree dies. Cambium feeders lay eggs in the living woody tissue, and the larvae hatch and usually eat enough that the tree can no longer transport nutrients and dies.

Delta method Modeling jargon for basing inferences on the differences between outcomes rather than the specifics of individual outcomes. For example, projections from GCMs usually focus on the future change of global temperature from the present (or past) rather than the actual value predicted.

Dendrochronology The study of trees (dendro) over time (chronology), usually via annual growth rings.

Dispersal kernel Mathematical term for the distribution of distances than a plant or animal might disperse. Typically these are right-skewed (in a population, there will be a lot of short dispersals and only a few long ones), and sometimes *fat-tailed*.

Dispersal vs migration One-directional movement (for animals usually out of a home range) vs. two-directional (out and back), either annually or once in a lifetime (e.g., anadromous fish).

Downscaling Jargon for translating coarse-scale observations or processes, in space or time or both, into finer-scale versions. It is most frequently applied to climate models' being downscaled from global to regional.

Early- vs. late-successional In the directional change of composition and structure of vegetation, the early stages, often after a disturbance, vs. the later stages.

Earth-system model A model of everything, suitably streamlined so that it does not have to replicate every process on Earth. A classic combination is atmosphere, oceans, vegetation, wildfire, land use, and fossil-fuel emissions, but there are many other possible combinations.

Ecohydrological model A simulation model that combines flowing water and the dynamics of ecosystems.

Ecological, evolutionary, geological time scales Time frames, from the shortest to the longest, associated with observable changes in these processes. For example, wildfires burn and trees grow in ecological time, bacteria and humans evolve in evolutionary time, and landforms arise and erode in geological time. So each covers several *orders of magnitude*.

Ecosystem services Things from ecosystems that benefit humans, often involving tradeoffs; for example, coal from strip mines vs. clean air and water.

Ecotone The transition between two more or less distinct types of vegetation, such as forest-to-meadow or shrub-to-grassland. The "alpine treeline ecotone" is of special interest to us in this book, and occupies the elevational band between continuous forest and alpine meadow.

Elevation-dependent warming Increasing temperatures at higher elevations within a region that are different—almost always greater—than the regional average.

Emergent In science, referring to a process or outcome that is qualitatively different from what might be expected from summing up more fundamental processes. A classic example is flock movement in birds caused by each bird responding only to its neighbor.

Endemic (1) Found only "there", as a species that is endemic to a region, or island. (2) opposite of "epidemic" when referring to insect outbreaks and other processes that overcome some local constraint.

Energy balance Measurement of the additions and subtractions of energy in a system, such as the Earth.

Energy-limited vs. water-limited A species, process, or ecosystem that would "do better" if energy or water were in larger supply. For example, a tree whose growth is curtailed by lack of sunlight vs. lack of water in the soil.

Ensemble In modeling jargon, repeated runs of one or more models with some parameters or inputs adjusted at the beginning or within to include uncertainties that are known or suspected. The outputs of the model will then be distributions rather than single values.

Equilibrium line altitude The dividing line on a glacier between the *accumulation* and *ablation* zones. This will change during the year with air temperature.

Evapotranspiration The combination of evaporation of water from a plant's surface and the direct expelling (transpiration) of water that has moved through leaves, stems, and flowers.

False precision In statistics and other quantitative sciences, and with models, this is an unwarranted narrowing of the range of an estimate, often from a logical error by the (human) estimator. Typically the estimate is of some metric or parameter, but it can also refer to the narrowing of a target, such as we saw in the example of a climate model: estimating the average temperature on a specific day when only a monthly average is warranted.

Fat-tailed Statistical jargon for a probability distribution in which there are more observations on the far right (i.e., greater) than "expected". "Expected" usually means a negative exponential distribution, although this is rarely stated. Distributions of extreme events are usually fat-tailed.

Fire suppression vs. exclusion In fire science, suppression refers to activities with the specific purpose of controlling or extinguishing fires, whereas exclusion refers to any process that leads to there being less fire somewhere. For example, aerial bombing and hosing are suppression; thinning a forest stand to make it less fire-prone is exclusion.

Flammability-limited vs. fuel-limited Referring to fire regimes in which there is always sufficient fuel but weather is not always conducive to fire vs. regimes in which it is almost always hot and dry enough to burn but there is not always enough fuel to sustain a fire.

Forcing Jargon for anything that makes something else happen. For example, increasing atmospheric CO_2 is a climate forcing.

Forward vs inverse modeling Basically, the difference between "What will this do?" and "What did this?" Forward modeling moves "forward" from causes to effects, whereas inverse modeling works backward from effects to causes.

Galleries In entomology (study of insects), nests for larvae within the live woody tissue of trees.

Gradient A more or less smooth change in one variable in parallel with another. For example, there is a temperature gradient associated with elevation: lower with higher except in cases like *cold-air drainages*.

Greenhouse effect Refers to the trapping in the lower atmosphere of long-wave radiation from the Earth, reducing its loss into space.

Greenhouse gases Gaseous elements and molecules in the atmosphere that prevent long-wave radiation from escaping. CO_2, methane, and water vapor are important ones.

Host species The species that "receives" (not always welcomes) another in some interaction, running from symbiotic to lethal. For example, bark beetles have preferred host species, as do insect parasites and some woodpeckers.

Hydroclimate The aspects of climate associated with water ("hydro").

Hydrograph A graph that uses data from one or more locations along a moving water body to display changes over time, usually daily, in the rate of flow.

Jet streams The largest moving air masses in the atmosphere.

Krummholz A stunted (not upright) growth form of trees, usually found at the upper elevational limit of a species.

Landscape resistance A measure of how hard it is for an organism to move across a landscape. For the same landscape, this can be very different for different organisms, and one of two landscapes can be more or less resistant than the other, depending on who is moving.

Life-history strategies Evolved patterns of growth, reproduction, predation, hibernation, etc., of plants and animals that reflect evolutionary and environmental pressures.

Limiting factors Anything that holds back the success of an organism. Spring snowpack is a limiting factor for growth of high-elevation trees; road density is a limiting factor for survival of grizzly bears.

Macrofossil Any piece of a fossilized plant or animal that is large enough to be recognizable (as opposed to deduced from DNA, for example).

Maritime vs. continental With climate, moderation of extremes by delivery of sufficient moisture by offshore air, vs. lack of the same, from being too far inland or blocked by a *rain shadow*.

Mass balance Measurement of the additions and subtractions of mass in a system, often (for us) ice mass in a glacier.

Mass movement Landslide or avalanche.

Matrix In landscape ecology, the "background", or surrounding part of the landscape, in contrast with patches, which are often the objects of active interest. One can also study the matrix, of course.

Megafauna Big (mega) animals (fauna). This usually refers to the biggest in a particular place, often extinct ones.

Mesic Referring to the middle ground between wet ("hydric") and dry ("xeric") in plant-water relations. A "mesic" species is one that prefers moderate moisture, which is characteristic of many mountain forests in the West.

Meso-predator A medium-sized animal (usually used for mammals) that is still big and fierce enough to be a predator but not a "top" predator. Foxes, raccoons, coyotes, fishers, and ermine are meso-predators.

Metric A way of representing something of interest numerically, often a combination of raw values. For example, we have seen water-balance deficit, degree-days, snow-water equivalent.

Microclimate Climate at the smallest scale, usually something distinct and often in contrast with the surrounding climate. For example, a tree clump in a subalpine meadow has its own microclimate - less variable than the surroundings.

Milankovich cycles Multi-millenial cycles in Earth's orbit: eccentricity (100,000 years—how far is the orbit from circular), obliquity (41,000 years—how much Earth's axis tilts), and precession (26,000 years—where Earth's axis points).

Monospecific Having only one (mono) species.

Montane A loose term referring to elevations (in the mountains) between those that are near lower or upper treelines.

Non-vascular In plants, those that have no circulatory system for water, and therefore are more sensitive to drought than those that do. Bryophytes (mosses and liverworts) and lichens are non-vascular.

Nonlinear dynamical system A mathematical representation of physical or biological processes in which responses to *forcings* are "out of proportion" to the forcing, either much larger than expected in a "well behaved" system or much smaller. Chaotic systems are a set of these that are so out of proportion as to be essentially unpredictable.

Ocean mixing In climate science, the rate at which heat in the upper ocean is transported to the depths. Estimating *climate sensitivity* depends on getting ocean mixing right.

Orders of magnitude Differences by factors of ten. For example, 10,000 is three orders of magnitude larger than 10. Most comparisons are not exact; 9766 is still three orders larger than 11.2.

Orographic A meteorological effect caused by topography. A *rain shadow* is an orographic effect.

Outbreak Way more of something fairly quickly, such as an outbreak of bark beetles, or measles, or political extremism.

Packing ratio In fire behavior, the packing ratio measures how dense fuel is, and therefore how much space there is for oxygen to support combustion. The "optimal packing ratio" is the one that enables the most intense fire. At the extremes, such as a huge block of wood (high) or a wide scattering of sticks (low), combustion is difficult or impossible. Good kindling is near the optimal.

Patch In landscape ecology, a piece of ground of particular interest, often embedded in a *matrix* along with other patches. Where patches are relative to each other, how big they are, and how different they are from each other and from the matrix are all of frequent interest.

Peak runoff The largest volume of water per unit time measured by a hydrograph, or calculated in a model.

Phase partitioning The calculated proportions of the two opposite sides of a *phase transition*, for us usually how much precipitation falls as rain vs snow.

Phase transition An abrupt (sometimes in time but mainly in character) change from one state to another, such as from ice to water or the reverse. In ecosystem, transitions are rarely so abrupt as to be proper phase transitions, but see Chapter 8.

Phenology The study of annual and other periodic changes in the life cycles of plants and animals, in response to climate and other environmental forcings such as the length of days. Important phenological elements of deciduous plants are the timing of loss of leaves in autumn and new growth in spring.

Pokilothermal vs homeothermal In animals, "cold-blooded" vs "warm-blooded", unable vs. able to regulate one's body temperature internally.

Population dynamics The study of populations, usually of animals, in the aggregate, as opposed to as individuals. Often highly mathematical.

Process-based Jargon for a model or a type of knowledge that has a good understanding of cause-and-effect, as opposed to "empirically based". There is much gray area (see text).

Proxy records Observations that substitute for others that we can't make. For example, tree rings are a proxy for temperature (and other things), and ratios of oxygen isotopes are a proxy for atmospheric CO_2 concentrations.

Rain on snow Just that. It snowed then it rained, causing the snow to melt faster than it would at the same temperature if dry. Winter floods are a common result of rain-on-snow events.

Rain shadow An effect, or a place, where something has blocked (created a shadow) the movement of moisture that would have brought rain. Usually this is the leeward side of mountains.

Realized vs fundamental niche Jargon for where a species actually is vs. where it could be if it were placed there. Usually a realized niche is taken to be a subset of the fundamental niche, i.e., a species that happens to be where it isn't expected to survive is not in a realized niche, but just in a temporary refuge.

Record length The temporal coverage of data.

Refugia Literally, refuges from an otherwise hostile environment, such as a stand of mature forest that escaped wildfire in a landscape of young trees.

Regional climate models Climate models whose domains are sub-continental to continental, as opposed to global. Regional climate models can represent detailed physics better than GCMs, because they have finer resolution (also called "grid-spacing" by researchers).

Respiration Breathing. Different in plants because they take in CO_2.

Rock glaciers Moving mixes of rock and ice. They can be either frozen rock debris or ice glaciers covered by fields of small-to-large boulders.

Serotinous Cones (or the species with them) that open only under intense heat, an adaptation to wildfire.

Shade tolerance How well a plant grows and survives under low light.

Simulation model Any model that replicates (and abstracts) reality to ask "What if?"

Sky islands Isolated mountain ecosystems in the Southwest or Great Basin that are like islands in that they lack *connectivity* with their surroundings.

Snow water equivalent The depth of water that would be produced by a volume of snow if it all melted.

Source vs sink Whether some structure or process releases more of some substance than it takes in (source) or takes in more than it releases (sink). A source has a negative balance, and a sink has a positive balance.

Species, genera, etc. vs. life forms Formal taxonomy vs. a qualitative way of distinguishing organisms, usually plants, by morphology. As with taxonomy, life-form classes can be hierarchical. For example, the first element of "tree, shrub, grass, forb, non-vascular" divides into "needleleaf, broadleaf".

Stand reconstruction Any method that works backward in time from observations to estimate past ages, sizes, composition, or density of plants, usually trees.

Statistical model A mathematical model that uses aggregate properties of collections of observations to draw inferences about the collection or parts of it. Correlation is one of the simplest statistical models.

Stepping stones Jargon used in studies of animal movements. A stepping stone is a place of temporary repose, such as a true stepping stone in the middle of a stream, or shade from trees in a hot treeless series of switchbacks (for a hiker).

Stochastic vs. deterministic Properties of models, whether *forward* or *inverse*. In a deterministic model, the pieces are specified precisely; in a stochastic model, some are varied intentionally, via probabilities, to compensate for imperfect knowledge of the system. Statements like "Fire is a stochastic process" are misleading. Fire does what it does, which is not throwing dice. We use stochastic models to represent fire because we have imperfect knowledge of that "doing".

Succession The change over time in the species of plants in an ecosystem.

Surface albedo A measure of how much the Earth's surface reflects incoming radiation, from low albedo (not much, as with forest canopies) to high (a lot, as with snow or ice).

Surging glacier A glacier that advances quickly, often without an obvious cause.

Symbiotic Literally life (biotic) next to (sym). Processes or organisms that use each other by virtue of being compatible and adjacent. For example, lichens are a symbiosis of algae and fungus.

Synergistic vs antagonistic "Together" vs. "against". Synergistic processes create an outcome that is more than additive (the sum of its parts). Antagonistic processes neutralize each other, partially or totally.

Synoptic In meteorology, the scale of large-scale phenomena such as cyclones. Between the global scale and the "mesoscale" (e.g., thunderstorms or tornadoes).

Teleconnection An association between weather patterns that are far enough apart to seem unrelated. For example, the El Niño Southern Oscillation (sea-surface temperature in the tropical Pacific Ocean) affects weather in the United States.

Thermodynamics vs. fluid dynamics Physical forces related to heat vs. fluid motion. Both influence climate and can affect each other. For example, temperature differentials in the atmosphere cause circulation, and vice versa.

Transverse In geography, side-to-side, as in the Transverse Ranges of California.

Treelines An often ill-defined edge between forest (or savanna or woodland) and non-forest. In most of the Western Mountain ranges there are upper and lower treelines.

Tropopause The boundary between the troposphere and the stratosphere.

Ungulate A hoofed mammal, of which deer, elk, bison, and wild sheep are western members.

Vagile vs. sessile vs. volant Moving (animals) vs. non-moving (plants) vs. flying (birds).

Voltine Referring to how many times a year an organism reproduces. In our example, beetles that reproduce less often than once a year are "semi-voltine", and those that reproduce once a year are "uni-voltine".

Water-balance deficit A *metric* representing how much less moisture is available to organisms than would be optimal. This has proven useful in estimating annual burned area at large scales.

Weighted average An average that gives unequal importance to its components. Suppose two dogs weigh 60 and 50 lb, and four cats weigh 12, 8, 7, and 11 lb. The average weight of a "pet" is (60 + 50 + 12 + 8 + 7 + 11)/6 = 148/6 = 24.67 lb. But what if dogs were twice as important (Cat lovers: you could do this with cats too, with reverse logic), because you might want to estimate the total effort of lifting these guys, and the dogs might hurt your back? So you "weight" dogs doubly. The weighted average is 60 × 2 + 50 × 2 + 12 + 8 + 7 + 11)/8 = 258/8 = 32.25. From this estimate, you are doing 33% more work than with the other.

Wildland-urban interface Where human structures, usually houses, and open space (i.e., not farmland) meet. It can be a straight line or very jagged. Either way, it creates big problems for firefighting.

Windthrow, or blowdown An area of trees that have fallen from being struck by wind.

Index

A

Abiotic, 109
Ablation zone, 70
Abstraction and complexity, 179–181
Accumulation zone, 70
Adaptation, 20, 21, 87, 91, 105, 111, 131–133, 136, 161, 179, 184, 198–207, 210, 218
Adaptive seasonality, 125, 141
Adjustable parameters, 56
Aerosols, 8, 30, 42, 46, 47, 49, 54–56, 59, 127
Age structure, 88
Air quality, 115, 187, 193–195, 197, 199, 208, 209, 213
Albedo, 8, 47, 59, 100, 101, 127
Algorithm, 67, 78, 105, 139
American Rockies, 15, 27–32
Anadromous, 149, 160
Anecdotal knowledge, 176, 201
Anticyclone, 24
Anti-fragile, 206–207
Assisted dispersal, 7, 94, 157, 200, 210
Atmospheric rivers, 66, 128, 170

B

Basin Ranges, 38–39, 41, 60–62, 108, 118, 140, 143, 166, 185, 186, 190–191
Bell-shaped curve, 12, 94
Bioclimatic envelope models, 104, 105
Biomass burning, 8, 47, 97
Biotic, 109, 149, 185
Boreal forests, 109, 118
Boundary conditions, 58, 79, 80
Business-as-usual, 186
Butterfly effect, 55

C

Cambrian explosion, 145
Capacitors, 74, 81, 184
Carbon cycle, 97, 195
Carbon sequestration, 167, 193–195, 197–199
Cascade Range, 15–29, 33, 60, 61, 66, 68, 83–86, 91, 92, 96, 106, 111–113, 118, 129, 130, 140–143, 152, 156, 158, 163–164, 174, 182, 185–186, 190, 200, 215, 216
Cenozoic, 16, 52, 70, 146
Change detection, 175, 176
Chaotic dynamics, 45, 183
Chinook, 130, 141, 160
Climate-change drought, 65
Climate-change event, 169
Climate forcing, 46, 47, 49, 50, 55, 56, 59, 99, 100, 186
Climate sensitivity, 47
Climate velocity, 94–96
Climatic envelope, 10, 21
Coarse-graining problem, 183
CO_2 fertilization, 47, 100
Cold-air drainage, 25, 58, 89, 182, 188
Colorado Plateau, 16
Columbia Basin, 16
Columbia Plateau, 16
Complacent, 81, 106
Conceptual models, 174, 177, 178, 181
Coniferous, 20, 83, 84
Connectivity, 10, 27, 33, 43, 149–151, 153, 185, 188, 199, 202, 212
Conservation biology, 207
Contagious, 110
Convective storms, 66